AF838208

THE GULF OIL AND GAS SECTOR

POTENTIAL AND CONSTRAINTS

THE GULF OIL AND GAS SECTOR

POTENTIAL AND CONSTRAINTS

THE EMIRATES CENTER FOR STRATEGIC STUDIES AND RESEARCH

The Emirates Center for Strategic Studies and Research

The Emirates Center for Strategic Studies and Research (ECSSR) is an independent research institution dedicated to the promotion of professional studies and educational excellence in the UAE, the Gulf and the Arab world. Since its establishment in Abu Dhabi in 1994, ECSSR has served as a focal point for scholarship on political, economic and social matters. Indeed, ECSSR is at the forefront of analysis and commentary on Arab affairs.

The Center seeks to provide a forum for the scholarly exchange of ideas by hosting conferences and symposia, organizing workshops, sponsoring a lecture series and publishing original and translated books and research papers. ECSSR also has an active fellowship and grant program for the writing of scholarly books and for the translation into Arabic of work relevant to the Center's mission. Moreover, ECSSR has a large library including rare and specialized holdings, and a state-of-the-art technology center, which has developed an award-winning website that is a unique and comprehensive source of information on the Gulf.

Through these and other activities, ECSSR aspires to engage in mutually beneficial professional endeavors with comparable institutions worldwide, and to contribute to the general educational and scientific development of the UAE.

The views expressed in this book do not necessarily reflect those of the ECSSR.

First published in 2006 by
The Emirates Center for Strategic Studies and Research
PO Box 4567, Abu Dhabi, United Arab Emirates

E-mail: pubdis@ecssr.ae
Website: http://www.ecssr.ae

Distributed by The Emirates Center for Strategic Studies and Research

ISBN 9948-00-809-X hardback edition
ISBN 9948-00-808-1 paperback edition

CONTENTS

Figures and Tables

Figures

TABLES

APPENDICES

Abbreviations and Acronyms

ABB	Asea Brown Boveri
ADGAS	Abu Dhabi Gas Liquefaction Limited
ADNOC	Abu Dhabi National Oil Company
API	American Petroleum Institute
ASPO	Association for the Study of Peak Oil and Gas
bcm	billion cubic meters
BPC	Basrah Petroleum Company
capex	capital expenditure
CCGT	combined cycle gas turbine
CCS	carbon dioxide capture and storage
CGES	Centre for Global Energy Studies
CERA	Cambridge Energy Research Associates
CFP	Compagnie Française des Petroles
CFTC	Commodity Futures Trading Commission
CPA	Coalition Provisional Authority
CSR	corporate social responsibility
CSIS	Center for Strategic and International Studies
DC	developing country
DOD	Department of Defense
DOE	US Department of Energy
DPC	development and production contracts
DrKW	Dresdner Kleinwort Wasserstein
E&P	exploration and production
EEA	European Economic Area
EIA	Energy Information Administration
EOR	enhanced oil recovery
FDI	foreign direct investment

FSU	Former Soviet Union
Gb	billion barrels
GCC	Gulf Co-operation Council
GDP	gross domestic product
GTL	gas-to-liquids
GUSA	Gulf–South Asia
IMOO	Iraqi Ministry of Oil
INOC	Iraqi National Oil Company
INTSOK	Norwegian Oil and Gas Partners
IDC	Iraqi Drilling Company
IEA	International Energy Agency
IOC	international oil company
IPCC	Inter-Governmental Panel on Climate Change
IPC	Iraqi Petroleum Company
kbpd	thousand barrels per day
KBR	Kellogg, Brown and Root
KOC	Kuwait Oil Company
KPC	Kuwait Petroleum Corporation
LNG	liquefied natural gas
LPG	liquefied petroleum gas
mbpd	million barrels per day
BTU	British Thermal Units
MEES	Middle East Economic Survey
MGS	Master Gas System
MPC	Mosul Petroleum Company
mpg	miles per gallon
Mt	million tons
NCS	Norwegian Continental Shelf
NGC	North Gas Company

NGL	natural gas liquids
NIOC	National Iranian Oil Company
NOC	national oil company
NOC	North Oil Company
NPC	Norwegian Petroleum Consultants
NPD	Norwegian Petroleum Directorate
OAPEC	Organization of Arab Petroleum Exporting Countries
OEC	Oil Exploration Company
OECD	Organization for Economic Cooperation and Development
OG21	Oil and Gas in the 21st Century
OPEC	Organization of Petroleum Exporting Countries
PDVSA	Petróleos de Venezuela, SA
PIW	Petroleum Intelligence Weekly
PMD	Project Management and Development
PPP	purchasing power parity
PSA	production sharing agreements
psi	pounds per square inch
ROACE	return on average capital employed
SADOP	Oil Products Distribution Company
SARA	Saudi Aramco Reserved Area
SCOP	State Company for Oil Projects
SEC	Securities & Exchange Commission
SGC	South Gas Company
SOC	South Oil Company
tcf/a	trillion cubic feet per annum
tcf	trillion cubic feet
toe	tons of oil equivalent
TSR	total shareholder return
UAE	United Arab Emirates

USACE	US Army Corps of Engineers
WMD	weapons of mass destruction
WEC	World Energy Council
WTI	West Texas Intermediate
WTO	World Trade Organization

FOREWORD

The global energy industry experienced a period of turbulence over the last few years with the Gulf oil and gas sector at the epicenter of this upheaval. The multiple risks posed by diminishing oil reserves, declining production from mature oilfields, and uncertainty over oil supplies from certain countries due to geopolitical upheavals have made supply security a matter of utmost concern for consumer nations. This supply insecurity, coupled with the burgeoning demand from the emerging economic powers of Asia pushed the price of oil to record levels.

OPEC members and Gulf oil producers in particular, are the obvious beneficiaries of increased oil revenues—at least in the short term. However, unrealistically high oil prices are economically unsustainable because they hamper global and regional growth, particularly in the OECD and developing economies. Moreover, very high oil prices could ultimately be counter productive for oil producers. The predominant position of oil in the energy slate could be undermined by a loss of market share as demand shifts away from oil to more affordable, cleaner and renewable forms of energy.

The current scenario poses special challenges for OPEC and the Gulf producers in particular. OPEC must formulate a cohesive long term strategy to regulate oil prices in a way that ensures reasonable production quotas and revenues for its member states and protects the organization's long term market share, while offering a reliable supply at an affordable price for consumer nations. In formulating and implementing policies for their future hydrocarbon development, the Gulf states must make important strategic decisions in terms of developing the most promising sectors and attracting the required levels of investment. The Gulf oil producers must strike the right balance between fulfilling their own strategic energy imperatives and the demands of the world economy.

The ECSSR Tenth Annual Energy Conference, *The Gulf Oil and Gas Sector: Potential and Constraints*, held in Abu Dhabi from September 26–27, 2004 examined the current position and future role of the Gulf oil and gas sector in this global scenario. Experts from the Gulf region and beyond discussed a range of pressing energy issues. OPEC's position and prospects came under the spotlight, with participants stressing the need for greater cohesion and cooperation within the organization. The relationship between Middle East reserves and production levels was examined through the prism of two regional case studies: Iraq and Saudi Arabia. The volatility in global oil prices, and its impact on supply and demand with all its global economic ramifications was analyzed. In this context, the Norwegian experience in using petroleum revenues to promote economic diversification and cushion the economy from oil price vagaries was also studied. The need to attract higher levels of investment in the Gulf energy sector and the potential for boosting inter-regional and intra-regional gas trade was emphasized. Recommendations were made for the effective management of the Gulf oil and gas sector to ensure optimum utilization of natural resources, achieve national growth objectives and support economic diversification goals while satisfying global energy requirements.

The ECSSR takes this opportunity to express its gratitude to all the conference speakers for their active participation and for sharing their insight through the publication of their papers in this volume. The Center would also like to thank the panel of expert referees who reviewed these conference presentations and offered their valuable comments. A word of thanks is also due to ECSSR editor Mary Abraham for coordinating the publication of this book.

Jamal S. Al-Suwaidi, Ph.D.
Director General
ECSSR

INTRODUCTION

The Gulf Oil and Gas Sector in the Global Energy Context: An Overview

The Gulf's oil and gas sector will continue to play a dominant role in the global market, as long as oil maintains its position as the world's main source of energy, and the natural gas industry continues to register record growth, buoyed by environmental factors. With global demand expected to reach new highs, oil production peaking in several areas and geopolitical crises heightening uncertainty, issues of supply security are raising serious global concern. The surge in petroleum prices witnessed during 2004 has brought a revenue windfall for the oil producers but threatens to clip global economic growth and impose a heavy burden on consumers.

In such an energy scenario, the Gulf oil and gas sector faces a crucial question: How can it meet its long-term strategic needs while adequately meeting global energy needs? Experts attending the ECSSR Tenth Annual Energy Conference *The Gulf Oil and Gas Sector: Potential and Constraints*, held in September 2004 examined both the possibilities and the problems faced by the Gulf energy sector in the context of global developments.

The conference presentations compiled in this book cover a range of important energy topics including the future of OPEC, the impact of world oil prices on global growth, the strategic objectives of the Gulf energy sector, the influence of markets and technology on regional reserves, new partnership models for national and international oil companies, the need and the prerequisites for boosting Gulf energy investment and the impact of new technological developments on the regional gas trade. Country studies outline the prospects for Iraqi oil production, highlight the policy options for Saudi Arabia as a swing producer and draw lessons from the Norwegian experience of economic diversification. The salient points of these conference presentations are summarized below to offer a broad overview of the prospects for the Gulf oil and gas sector in the light of global energy developments.

Gulf Oil and Gas: Potential and Constraints

The proven oil and gas reserves in the Gulf region are considered the largest in the world. With the projected increase in world energy consumption, the region's future role will be to provide much of the hydrocarbons required to satisfy this global demand. H.E. Abdullah bin Hamad Al-Attiyah points out in his Keynote Address that much needed hydrocarbon reserve potential is available in the region. What has been achieved in the Gulf oil and gas sector so far, ensures sustained production growth, industrialization and energy exports.

The main objective for the region is to achieve growth and expansion, not only in quantity, but also in maximum efficiency, to guarantee the optimal utilization of natural resources and investments in processing these resources. In order to achieve integrated and optimal utilization of regional natural resources, there must be intensive effort and investment in the development of material, human and social capital.

Human resource development, including leadership training in all fields is imperative for sustained and lasting development. Technology transfer cannot be effective if intellectual property rights and the latest products are merely purchased from others. It involves having trained national personnel capable of handling and applying technology and tailoring it to the region's needs and purposes. Another fundamental task is to develop material infrastructure and basic services, which will considerably reduce the investments necessary, shorten the time span required for various projects, as well as realize considerable cost savings in the general development process.

Undoubtedly, there are wide horizons for regional cooperation. There are many cases in which the fruits of joint ventures and coordination have not fulfilled regional expectations. It is necessary to coordinate regional economic plans and fields of industrialization to limit duplication and competition in local and global markets and benefit from the economics of large-scale projects, as well as to make use of contractual experiences and expertise in the best interest of investors in the region, whether government, mixed sector, or even private sector. In a world that is increasingly attuned to globalization, the region must develop systems of work and investment that meet specific business needs but are compatible with the regulations governing economic and industrial development under which the region will have to compete sooner or later.

The Keynote Address concludes that the oil and gas sector in the Gulf possesses enormous potential, which can be expanded through regional cooperation and that this sector can prosper even more if existing constraints are overcome.

The Future of OPEC

Since its inception, the petroleum industry has always been global in character, observes Adnan Shihab-Eldin, and over the years it has adapted to changes and grown accustomed to dealing with global risks and benefits. However, the accelerated pace and strength of globalization forces make its effects on the industry more far-reaching and pervasive, and present it with new opportunities as well as formidable challenges. In this situation, the role of the Organization for Petroleum Exporting Countries (OPEC) in coordinating oil supply and regulating oil prices deserves particular attention.

Notwithstanding the unusually high oil price levels observed in 2004, OPEC, in line with its core objectives of maintaining stable markets and reasonable prices, has succeeded in ensuring that the market remains well supplied with crude throughout the year and that the fundamentals are sound. The rising price trend of 2004 was attributable to the exceptional confluence of many factors, including higher-than-expected oil demand growth, downstream and distribution bottlenecks and geopolitical tensions. An examination of the short-term oil market reveals the centrality of oil in meeting mankind's energy needs, its dynamic, constantly changing nature and its inherent volatility. At the same time, it underlines OPEC's important role in seeking to regulate and stabilize the market.

Over the longer term, the next two decades are expected to see fossil fuels continue to account predominantly for increases in energy demand, with oil maintaining its major role as a source of energy. There is also a clear expectation that there is a sufficiently abundant oil resource base to satisfy this demand growth. Three contrasting scenarios are presented for the period up to 2020: dynamics as usual, high growth and low growth. The ranges of demand and supply growth and absolute magnitudes throughout the period are explored, underlining the profound difficulties facing producers as they seek to meet ongoing challenges such as the capacity expansion and spare capacity management of OPEC member countries. Although non-OPEC production is viewed as expanding over the medium term, it is generally agreed that OPEC will gradually be relied upon to supply the incremental barrel, with its market share eventually set to rise. In this outlook, the investment required by the oil industry in absolute terms, is not necessarily different from that observed in the past.

Maintaining stable prices will be complicated by future uncertainties as to the evolution of demand and the level of non-OPEC supply. Uncertainties over future economic growth, policies and the rate of development and diffusion of newer technologies, raise questions over the required scale of investment. This uncertainty, coupled with long lead times, inevitably complicates the task of maintaining market stability. Additionally, medium-term prospects suggest there is a need to ensure that the level of spare capacity is consistent with such stability. The key to ensuring that sufficient investment takes place is oil market stability, whereby prices are maintained at reasonable levels to secure adequate sources of investment.

Technology development and the future of oil are examined from the perspectives of both supply and demand. Technology can contribute to demand growth, just as it can reduce demand through improved efficiency. Given that fossil fuels are set to provide most of the increase in energy demand over the next two decades, there is a need for cleaner oil technology. Sustainable patterns need to take advantage of promising technologies, such as carbon sequestration, reductions in gas-flaring, and clean oil in electricity generation. The commercial potential of CO_2 sequestration, combined with enhanced oil recovery (EOR), presents one of the opportunities for cooperation on technological issues, with the promise of great benefits for all.

Maintaining OPEC Cohesion

While acknowledging the significance of OPEC's efforts to coordinate oil supply and stabilize prices, Luis E. Giusti underlines the need for greater internal cohesion if the organization is to enhance its effectiveness in regulating the global energy market. The birth of the organization in September 1960 was a direct consequence of the short-sightedness of the international oil companies. At that time, the founding group was the source of 80% of crude oil exports and the primary objective of these oil producing nations was to defend the oil price. The founding members also called for a system regulating production, pledged solidarity in case the oil companies sought to impose sanctions on any member, and vowed to ensure stabilization of prices, eliminating harmful fluctuations.

Thus the creation of OPEC was a very important first step taken by the producing countries, but difficult times lay ahead, because the oil companies still retained control over market access. It was not until late 1973, that OPEC assumed control over the regulation of oil prices, independently of the oil companies. Another important step in the evolution was the process of nationalization in most OPEC countries, which transformed the oil companies into mere lifters of OPEC crude Paradoxically, those events translated into market volatility, because many OPEC decisions became increasingly influenced by external political events.

Over the years, OPEC lost market share, and when prices began to weaken, the organization decided in March 1983 to engage in a system of supply management, implementing a mechanism of ceiling and quotas. With that step, OPEC assumed the *de facto* role of supplier of the last resort, also assuming, unofficially, the responsibility for balancing supply and demand. However, some experts consider this system to be inherently flawed, because in contrast to the oil companies, OPEC has no access to the latest data about stock movements and stock build-up at the levels desired by refiners, market requirements of products and the seasonal adjustments, as such information is generally held by the oil companies. Therefore, OPEC's supply management tends to be deficient, and becomes a potential source of price volatility. After four and a half decades it can be said without hesitation, that OPEC has achieved reasonable success. However, the vision, direction and cohesion of the organization as a group remain weak. To establish a common and cohesive interest, as evidenced in 1960, is easier said than done. Differences among members have surfaced over time as the particular needs and concerns of each member state are often difficult to integrate into effective unified action.

In assessing the future of OPEC, there are important reflections to be made and crucial questions to be answered. What are the strategic issues confronting OPEC in today's world? Who and where are the "foes"? What is the right price target, and what are the considerations that should guide the definition of this target? Is the current price band appropriate? Is there a reasonable way for OPEC countries to balance their fiscal needs with price and volume objectives? How should the dilemma of market share versus price be resolved? What are the expectations of individual members concerning their quotas? Under what price conditions would substitution begin to threaten oil as a source of energy? All these questions must be addressed as OPEC cannot afford the luxury of having an unclear vision or erring in its strategic choices.

Iraqi Oil Scenarios for the Future

The continuing uncertainty over Iraqi oil supply has not only posed special challenges for OPEC but also cast a lingering shadow over the global oil market. The US-led invasion of Iraq in 2003 had initially raised hopes that the country's massive reserves could be tapped. Many countries eyed the possibility of benefiting, one way or another, from its oil riches. Oil companies with high expectations, were eager to participate in the development and exploitation of the country's ample and low-cost oil reserves. Oil analysts started forecasting optimistic scenarios of rising Iraqi oil production.

However, by 2005, more than two years after the collapse of Saddam's regime, none of these expectations seemed likely to materialize within the time frame initially envisaged. Indeed, Iraq has failed to deliver. Muhammad-Ali Zainy analyzes developments within the country, presents likely scenarios and outlines the necessary conditions for Iraq to regain control over its oil facilities and expand production to levels commensurate with its oil reserves.

Although Iraq is a member of OPEC, the organization does not presently bind Iraq by its quota system. In consideration of Iraq's special circumstances, OPEC kept Iraq out of the quota system since the latter resumed oil exports in accordance with the Oil-for-Food Program, which became effective in December 1996. This special treatment continues and may continue for the foreseeable future.

The critical factor presently hindering Iraq from increasing its production is its dire security situation. As conditions gradually return to normal and become safe for investors, credit institutions and contractors of all kinds, the prospect of the country reaching a sustainable production capacity of 6 million bpd by 2010 will become very real. With such a potential rise in Iraqi production in the foreseeable future, two legitimate questions will arise:

- Can the world oil market accommodate Iraq's new, and rather large, oil supplies?
- With the expected rise in Iraqi oil production, will OPEC continue to give Iraq special treatment and let it produce freely without being quota-bound? Or will it include Iraq within its quota system and hence constrain its production?

There is also the issue of what should be done with Iraq's enormous potential oil wealth. Several scenarios have been put forward, including the possible privatization of the country's oil industry. While it is quite feasible to privatize downstream operations, the same is not true of upstream oil operations, especially development and production, since all the rent resides in that side of the industry. However, a national Iraqi oil company, run efficiently by qualified personnel can capture all the rent that ought naturally to belong to the Iraqi people.

In the event of privatization, what would Iraq do with the sudden influx of potentially massive oil wealth? Certainly, the country will need tens of billions of dollars for reconstruction and development. Should the necessary finance for such activities, either from donor countries or from oil revenues, come gradually as needed, because Iraq's absorptive capacity for investment capital is limited? Indeed, one of the top priorities now is to win forgiveness of Iraq's foreign debts to the maximum extent possible. Without such a debt cancellation effort preceding the sale of some Iraqi oilfields, foreign creditors will be swarming around the suddenly cash-rich Iraq demanding payment of their mostly odious debt.

Another proposed scenario is to create an oil fund to even out fluctuating revenues emanating from oil price volatility, thereby stabilizing the government's income. Although this is a good proposition, it is not a panacea for all the country's ills, as successive governments can tamper with such funds.

Another interesting scenario is the direct distribution of a country's oil revenues to its people. This has the immediate effect of increasing the per capita income, but may cause inflation, especially if such income is largely spent on consumption. Currently, Iraq's productive base is decidedly too weak and inadequate to cater for the concomitant increase in demand.

There could be other viable scenarios for efficiently exploiting a country's oil revenues. The important thing, however, is that all of these scenarios are not mutually exclusive. Any combination of these can be adopted to suit the country's unique circumstances. In the case of oil-rich Iraq, the top priority would be to create a diversified, tax-based economy, with little dependence on oil. Nevertheless, so long as there are oil revenues, they should be exploited prudently for the benefit of the country and its people.

Saudi Arabia: Strategic Outlook and Options

Against a backdrop of supply uncertainty on the Iraqi front, Saudi Arabia's importance as the world's main oil supplier and swing producer remains undiminished. In Vera de Ladoucette's view, the country shows huge potential, including possible undiscovered reserves that may dwarf other "prolific" areas. Exploration effort so far has been limited and it is very likely that there are still some giant fields awaiting discovery. According to Saudi Aramco, the proven reserves are reported to be 263 billion barrels, with another 103 billion barrels of probable plus possible reserves. Saudi Arabia holds 21 percent of world proven reserves and provides some 12 percent of daily production. The much-publicized doubts about its field performance appear to be unfounded, since they are based on limited technical publications that generally focus on production problems and not their solutions.

With reserves expected to last 78 years at current production levels, Saudi Arabia's concern is not only to secure its market share today against non-OPEC supply, but also to preserve the predominance of oil in the global energy mix. This means ensuring that very high oil prices do not encourage investments in non-OPEC countries and do not stifle demand; taking a pro-active attitude to environment regulations; and closely monitoring the development of alternative energy sources in transportation—especially fuel cells.

Saudi Aramco continues to enjoy a monopoly on upstream oil. While policy-making and regulations are set by the Ministry of Petroleum and Mineral Resources, it is managed like any other international company. It focuses on efficiency and the in-house development/or the acquisition of the latest technologies. Saudi Aramco's challenge is to mobilize its own and cross-industry resources to promote the use of oil and gas. Carbon sequestration and capture technologies, as well as technologies connected with hydrogen generation, are being emphasized in its R&D programs.

Saudi Arabia is also seeking to secure outlets in its major markets by investing in their downstream sectors. Taking into account that the bulk of the demand growth is expected to keep coming from Asia, Saudi Arabia has shifted its downstream priorities to South Korea, Japan and China. To preserve its unique position in the oil market, the Kingdom has to maintain its usual 1.5 to 2 million bpd spare capacity cushion.

The main challenges confronting Saudi oil policymakers are: choosing the right time for the development of new fields; fine tuning OPEC production management so that oil prices do not, in the long run, affect demand and/or encourage non-OPEC supply; and making sure that Saudi Aramco's social responsibilities do not override sound investment decisions.

Saudi Arabia does not enjoy the same supremacy in the gas sector as it does in the oil sector. The country ranks fourth in terms of gas reserves and holds the eleventh position among the gas producers in the world. This has led the Kingdom to differentiate its gas approach from its oil policy. The recent focus on gas is driven by a surge in domestic demand, a growing need for basic utilities such as water and power and the need to develop an industrial base fuelled by gas, since gas-based utilities and industries account for 15 percent of GDP in Saudi Arabia.

In this context, Saudi Arabia has to take several measures: expand current proven reserves of 235 trillion feet of gas reserves through an aggressive exploration program; increase the share of non-associated gas to avoid boosting oil production and meet OPEC quota constraints; take confidence-building measures to convince the petrochemical industry that its feedstock needs will be met by adequate gas supply; address the concerns of the petrochemical industry about "dual pricing" for liquefied petroleum gas (LPG) for the local industry and for export; and allow foreign investments for gas exploration activities to free up public funding for social requirements.

Saudi Arabia's priority is to monetize gas by expanding its domestic market. However, at one point, it will have to face a challenge and decide whether to explore gas export opportunities by LNG or choose the gas-to-liquids (GTL) path or perhaps both.

The Norwegian Experience of Economic Diversification

Economic diversification is a recognized priority for the Gulf oil producers and Norway's experiment in petroleum-funded diversification is instructive. In the light of the country's background and its rationale for diversification, Øystein Noreng reviews the different policies Norway has adopted: local procurement, transfer of knowledge and petroleum revenue investment. In the process, he highlights the significance of the Norwegian experience for the Arab Gulf states, despite the differences in their respective backgrounds. Some basic issues and questions that arise include the impact of government intervention versus reliance on market forces, the value of infant industry protection, the significance of the excessive dependence on unstable resource rents stifling growth in other sectors, and the eventual impact of institutions and policies.

To what extent did the various protectionist policies in relation to the petroleum industry, concerning local content, knowledge transfer and industrial employment, succeed in reducing dependence on the petroleum industry? To what extent did they cause the country to suffer from a bout of "Dutch disease" in disguise, in the form of a resource curse intensified by institutions? Indeed, the local content policy made parts of Norwegian manufacturing exceedingly dependent on the local offshore market. The knowledge transfer policy concentrated the country's research efforts on petroleum-related matters. Significantly, as these policies of micro-economic intervention were abandoned in the mid-1990s, there was a general understanding that they had contributed to high costs in the petroleum industry as in other parts of the economy.

Since the mid-1990s, Norwegian policy has been to accumulate the petroleum revenues in an investment fund, the intention being to avoid harmful domestic inflationary pressures due to the expenditure of economic rent. So far, this policy of macro-economic management has been reasonably successful, but the risk is that huge surpluses on the current account balance will drive the exchange rate up, raising the domestic cost level and compromising the competitiveness of other industries, with the resource curse materializing through a financial endowment.

When it first embarked on its petroleum activities, Norway clearly benefited from the rule of law, a well-functioning civil service and a low level of corruption resulting in producer-friendly institutions. This framework attracted entrepreneurial talent to petroleum-related activities. The swift and successful implementation of protectionist measures signified, however, that the institutions also were grabber-friendly, underlined by the early use of anticipated petroleum revenues in the 1970s to support farmers, ship-owners and manufacturing in general. By the mid-1990s, the oversized oil supply and service industry indicated that Norway's institutional framework had some grabber-friendly aspects. At the present time, the export levels of the oil supply and service industry are growing, a sign of international competitiveness and evidence that Norway's institutions are once again producer-friendly.

Large resource revenues make the government more independent of a productive income base and free from taxes levied on civil society, usually leading to a consolidation of power, and often strengthening dictatorships in developing countries. Oil revenues form the foundation of several autocratic *rentier* states of the Middle East, where oil wealth is often viewed as a substitute for democracy and sometimes works as an impediment to institutional development. In industrial countries with established democracies, such as the Netherlands or Norway, large resource revenues to some extent decouple expenditure from productivity, enlarging the freedom of action in economic policy, making the political system a more tempting battle ground for interest groups, and encouraging arbitrariness and rent-seeking through selective measures. Norway is a case in point. The country has, so far, avoided the worst excesses of the resource curse, but huge petroleum revenues do carry substantial risks.

Another question is to what extent the reliance on market forces, with a less interventionist state in petroleum matters and a stronger position for foreign companies, would have produced comparable results in areas that have mattered to the Norwegian public, namely the integration of the petroleum industry into Norwegian society through jobs, deliveries and national control. This is a complex issue, not limited to monetary costs and benefits, but also to qualitative aspects of Norwegian society that are difficult to quantify, let alone monetize.

World Oil Prices and the Impact on Economic Growth

The oil price remains crucial for Gulf producers who may seek to bankroll economic diversification through oil-generated revenues. However, the future course of oil prices also has serious implications for the healthy growth of the global economy. While forecasting oil prices has always proved difficult, Mehdi Varzi identifies a number of key indicators:

First, oil price expectations are definitely on the rise. In the 1980s to mid 1990s, any price above $25 per barrel was viewed as being too high and unsustainable over the medium to long term. After OPEC successfully targeted a price band of $22–28 per barrel, market expectations rose gradually to the $25–30 per barrel level. Today, some hedge funds indicate that prices could remain above $45 per barrel for some time and even go higher than the current level of around $55 per barrel. Although not necessarily an accurate forecasting instrument, WTI crude futures stretching out to 2010 currently show prices at over $45 per barrel.

Second, with the world economy and world demand likely to continue on a growth path beyond 2005, there is unlikely to be a sharp drop in the demand for OPEC crude. Non-OPEC supply at best is likely to match rather than exceed annual incremental demand growth, leaving the demand for OPEC crude in the region of 28–30 mbpd.

Third, although OPEC capacity is set to rise, it is unlikely to result in a flood of new oil overnight. If OPEC members fail to re-establish a more comfortable level of spare capacity, the world oil market will continue to suffer from concerns about supply stability. Today, OPEC is virtually producing at full capacity. Thus an interruption of 2–3 mbpd would send shockwaves throughout the oil market. It would also result in a more negative global economic environment. In the event of a major oil price shock, if prices went back to the 1979 levels, a global recession cannot be ruled out altogether.

Why does OPEC capacity always seem to rise more slowly than anticipated? Some of the reasons include: domestic imperatives; capital constraints; technology gap; going for the most easily accessible oil without due regard for longer-term factors; and the lack of incentives for employees. The general conclusion is that these state oil companies have tended to act rather like the government bureaucracy, and have imparted little dynamism to stimulate upstream activity.

In Varzi's view, OPEC members have tended to focus too much on prices rather than on prospects for future world oil supply. The organization's greatest long-term enemy is the growing global concern being voiced about supply reliability. As a result of the price hike of 1979–1981, there was a wave of energy conservation and oil substitution measures by the major oil-importing countries. Such measures could well be implemented again if supply stability tops the agenda in Western capitals and in countries such as China and India. Forthcoming OPEC meetings should address this issue in a coordinated fashion and on an urgent basis.

Moreover, there is a need for greater transparency in OPEC's provision of energy data. While OPEC members have become more open than in the past about their production numbers, there are no reliable independent estimates of proven and probable reserves. As long as this is the case, certain observers, rightly or wrongly, will continue to insist that OPEC capacity is unlikely to rise much further and that the world is on the verge of a catastrophic global oil shortage. At the very least, greater transparency on upstream data and future capacity expansion would reassure the world that, under the right conditions, OPEC would be able to raise its capacity to substantially higher levels. It would certainly help politicians and central bankers to make more informed judgments about the potential impact of global oil fundamentals on the prospects for the world economy in the coming years.

Market and Technology Influences on Regional Reserves

As energy consumption trends reveal the world's apparently insatiable thirst for oil, Kjell Aleklett and Colin J. Campbell sound a timely warning about the peaking of oil production in the near future and the diastrous impact that the depletion of this crucial commodity will have on the global economy.

The first half of the Oil Age lasted 150 years since the first wells were drilled on the shores of the Caspian and in Pennsylvania. The cheap and abundant supply of energy from oil facilitated the growth of industry, transport, trade, and agriculture, and led to the growth of large amounts of new financial capital, which in turn gave rise to the subject of Economics by which to manage investment, wealth and finance.

Available indications are that the second half of the Oil Age will dawn over the next few years. It will be characterized by the decline of oil production and all that depends upon it, including the prevailing financial structures. There are many different scenarios for modeling future production. The post-midpoint countries (which are past their peak and already firmly into decline) will continue to decline at their current depletion rate. Most of the pre-midpoint countries are now within a few years of midpoint, so the assumptions are not very critical to the longer term. On balance, it seems reasonable to assume that production to midpoint will be flat, based partly on the limits of installed facilities.

The most difficult challenge is to forecast future Middle East production. Previously, the assumption was that the region could, and would, adopt a swing role, making up the difference between world demand under alternative scenarios, and what the other countries can deliver under the model. However, this assumption no longer seems plausible following the invasion of Iraq. Although these countries have relatively low depletion rates, suggesting a physical capacity to increase production, it seems less likely that this will in fact happen for three reasons. First, most of the region's oil is contained in a small number of giant fields found long ago, which are now heading into decline. Second, the industry in these countries is mainly in state hands and lacks the indirect subsidy available to foreign companies, which are able to offset exploration and operating expense against taxable income. Third, these countries have no real incentive to make the massive investments necessary to increase production, which would serve only to reduce their revenues by lowering oil price, and deplete their reserves more quickly.

If the evaluation of oil producing countries is summed up, the peak of global production, as computed, arrives in 2005 for regular conventional oil and a few years later for all categories. However, this is no more than a shallow peak, so that relatively small changes in the estimates and modeling could shift it by a few years. Even so, the evidence suggests that, by 2010, terminal decline would not only have set in, but also be perceived to have done so.

It seems likely that an increasing number of producing countries will follow the example of the previous Iraqi regime and trade oil in Euros or other local currencies to benefit from some of the hidden tribute now enjoyed by the United States. If so, the strength of the dollar and eventually the world stock markets, which are dominated by US interests, are likely to be further undermined.

In short, it may be said that the decline of oil heralds the "end of economics" as currently understood. Its principles and practices successfully delivered economic growth in parallel with rising oil supply, but new methods and practices are needed to manage the long term decline which characterizes the second half of the Oil Age. Furthermore, political structures in so-called democratic countries have mainly aimed at fostering conventional economic practices, with particular emphasis on liberalizing markets. These may also have to change in the face of the new circumstances. Whatever the outcome and the uncertainties, it seems certain that the peak of oil supply represents a discontinuity of historic proportions. In fact it is unprecedented because never before has a critical commodity declined from natural depletion before a better substitute took its place.

The world desperately needs to make sensible plans to face this little understood situation. Importing countries could cut imports to match world depletion rate. This would moderate world oil prices, by putting demand in better balance with supply. It would allow the poor countries to afford minimal needs; avoid destabilizing financial flows; and, above all, force the consumers to face reality. It would also serve the Middle East by encouraging those countries to conserve the resources on which they so heavily depend. The first step in such a process is to secure more transparent information on what is the true depletion rate. If valid statistics entered the public domain, the nature and impact of depletion would be entirely self-evident.

The Future of Gas Transportation in the Middle East Region

Fears over oil depletion and ecological considerations have served to boost the fortunes of gas as a cleaner alternative fuel. Until recently, the high costs of gas transportation had practically excluded the Middle East from cross-border gas trade. In 1993, only 6.5 percent of Middle East gas production crossed an international border, and since 3.7 percent of this represented trade among the constituent members of the United Arab Emirates federation, only 2.7 percent was truly international—the export from Abu Dhabi to Japan of 3.4 billion cubic meters (bcm) of liquefied natural gas (LNG). However, that situation is changing rapidly. A decade later, in 2003, trade accounted for 15.2 percent of Middle East production and LNG exports had increased tenfold. The International Energy Agency (IEA) now expects that the Middle East will be the fastest growing source of gas for inter-regional markets over the next two decades.

James T. Jensen identifies three factors as being primarily responsible for boosting the outlook for gas exports: first, gas-fired combined cycle gas turbine (CCGT) technology has made gas the fuel of choice for power generation throughout the world; second, the rapid growth in gas demand has stimulated interest among previously gas-poor countries and taxed the availability of traditional North American and European gas supplies; and third, technological improvements in LNG and pipelining as well as gas-to-liquids (GTL) conversion have helped to bring down the costs significantly.

CCGT units have revolutionized the use of gas in power generation. The thermal efficiency of the conventional steam boiler is limited to about 38 percent. However, by adding a high-temperature gas turbine on the front end of the steam cycle, CCGT unit efficiencies approach 60 percent. CCGT units also have low capital cost, can be built with short lead times and come in "bite-sized" modules that are well-adapted to market growth. Moreover, in an environment of growing concern over carbon dioxide emissions, the CCGT's combination of high thermal efficiency with a high hydrogen-to-carbon ratio fuel is an added advantage.

The rapid growth in gas use for power generation has created demand in previously gas-poor countries, such as Spain, Turkey, India and China. In addition, it has stimulated gas demand growth in North America and Europe to the point where demand is taxing traditional sources of supply.

This is especially true for the United States and the United Kingdom. Demand continues to grow even as indigenous production in the United States and Canada has apparently leveled out and the United Kingdom is facing a decline in North Sea production.

Simultaneously, there have been cost improvements in LNG, gas pipelining and GTL technology, making it possible to move natural gas over longer distances. The relatively lower cost of gas transportation coupled with increased interest in cross-border trade has positioned the Middle East as a major player in the international natural gas trade.

Pipelines, LNG and GTL all have a role to play in transporting gas out of the Gulf region. Pipelines are at their best for shorter hauls and thus should dominate local and regional trade. For intermediate distances, such as for the Indian subcontinent and Turkey, pipeline growth will be inhibited by politics and competition. India will probably prefer LNG for the nearer term although successful negotiations among India, Pakistan and Iran could change that conclusion. For longer distances, such as to Europe, pipelines may potentially be competitive, but they must be large and the intermediate country transit problem is a major barrier.

Increasing LNG demand in the distant markets of North America, Europe and Northeast Asia, coupled with the Middle East's large reserves guarantee rapid growth in LNG exports to these markets. GTL is the only challenger to LNG for very long hauls but it competes in a different ultimate market. The key to GTL's success is whether the product quality premium will be enough to offset its higher costs and lower thermal efficiency.

Investment in Middle East Oil: Role of IOCs and NOCs

Expanding the oil and gas infrastructure to increase production capacity and build pipelines and terminals will undoubtedly require high levels of regional investment. However, the effective mechanisms by which to achieve successful investment and business outcomes are not always clear. In this context, Valerie Marcel reviews the respective roles of the international oil companies (IOCs) and the national oil companies (NOCs) and suggests new options for building mutually beneficial industry partnerships. Both IOCs and NOCs profess interest in working together. In private, however, both groups have difficulty in finding common ground and often voice frustration about each other's expectations. Dissatisfaction on the part of the IOCs with previous joint ventures and increased confidence on the part of NOCs in their own capacity had led many on both sides to privately question the need for working with the other—at least under the present terms. Clearly, the current investment models do not allow both parties to align their respective interests.

Do IOCs and NOCs really need each other? IOCs want access to equity, acceptable rates of returns, incentives for enhanced recovery and opportunities for recurring investments. However, IOCs have equity access to only 14% of the world's oil and gas reserves in countries without NOCs, and to 11% of reserves in countries with NOCs. About 58% of oil and gas reserves are presently held by NOCs in countries where IOCs have only limited involvement through service contracts or technical service agreements. These terms, which are on offer in the large reserve countries, often do not meet the IOC criteria for investment.

Producers would like to see the oil majors increase their oil price assumptions to match that of OPEC. Indeed, bullish or conservative price assumptions have a direct impact on the profitability estimates of future investments. Nevertheless, a number of IOCs accept less than satisfactory terms of investment for small scale or short-term investments, in order to build relationships that they hope will give them access to reserves and better returns. Producers also want capital, management skills, and technology—with various countries ranking those priorities differently and some high competence NOCs, such as Saudi Aramco, expressing no such needs. Clearly each NOC is unique, but interviews conducted by the author with NOC managers and governments of Middle East oil producers indicate prevailing attitudes and emerging trends.

Investment capital is needed when the producers' fiscal relationship with the state is structured in such a way that their capital needs are sacrificed to government budgetary needs or when their means of revenue generation cannot meet investment requirements. Capital needs are expected to gain importance across the region because of the high rates of capacity utilization in a number of producing countries during 2003–04. Moreover, a slow-down of non-OPEC growth rates means that investment will be needed to increase capacity.

However, most large reserve holders in the Middle East will not give equity access to their reserves. Moreover, they will seek to maintain control over the management of their resources and maximize government returns. Agreement can more easily be reached on returns, but present investment models are not sufficiently flexible to address other concerns. Granting international oil companies the coveted right to "book reserves" allowed Algeria and Libya to attract much needed investment and kept investors highly committed in Abu Dhabi, but equity stakes in most other large reserve holders in the region are excluded.

Paradoxically, although the oil majors would seek opportunities to explore and develop new areas to beat the depletion game, they will in fact increasingly be called upon by the major reserve holders in the Middle East to assist them in managing their reservoirs and offsetting the decline of their mature fields. Projects in mature areas present bigger technological challenges and higher costs just to maintain production levels. NOCs are under similar pressures to invest in order to control depletion rates.

IOCs may not want to undertake this type of work. Indeed, if more significant IOC investment is not taking place in the Middle East, it is partially attributable to the reticence of the oil majors. Should they decide to work as contractors for an NOC or enter into partnerships with NOCs to develop existing fields, IOCs need to engage in a different kind of relationship with the NOCs and with the government. The IOCs would need to change their mindset, their corporate culture and be prepared to accept lower returns. IOCs also have to recognize that producing countries will naturally want to control their resources.

Successful new business models will depend on a careful alignment of each party's objectives, needs and assets. NOCs are unlikely to give turnkey contracts as they want to control and participate in development. This means that IOCs must either work for the NOC or work with it. By developing new business models, they can become partners. Successful partnerships will lead the NOC to gain new competencies and increasingly compete with the IOCs. Potential new options for NOCs and IOCs to join forces could range from making the IOCs integrated service providers for the producer client, to creating international strategic alliances throughout the supply chain.

Middle East Investment Requirements and Capabilities

Regardless of the future business models employed by NOCs and IOCs operating in the Gulf states, there is no question that massive investment will be needed to replace, build and sustain energy production capacity. The IEA's *World Energy Investment Outlook* concluded that worldwide investment requirements for energy-supply infrastructure over the period 2001–2030 amount to $16 trillion, or $550 billion a year. This investment is needed to replace existing and future supply facilities that will either be exhausted or become obsolete during the projection period. These funds are also necessary to expand supplies to meet the projected primary energy demand growth. Capital needs will grow steadily through the projection period. The average annual rate of investment is projected to rise from around $450 billion in the current decade to $630 billion in the 2020s. Although the total investment needs are large in absolute terms, they are modest in relation to the size of the world economy, amounting to only about 1% of global GDP on average over the next thirty years. However, needs differ sharply among regions, ranging from half a percent of GDP in the OECD countries to 5% in Russia.

Fatih Birol identifies several factors that could discourage, prevent or delay investment in particular regions or sectors. Prices will have to be high and stable enough to ensure adequate returns. Access to reserves, fiscal and commercial terms, the evolution of environmental policies and technological developments will also be key factors in future investment. There is considerable uncertainty about the amount of energy resources in certain regions and about how much it will cost to exploit them.

Given its huge resource base, the level of investment in expanding oil and gas production capacity in the Middle East will be vital to global energy market prospects in the medium to long term. Mobilizing that investment will depend largely on the production and investment policies of the key producers. Although the costs of developing the region's vast reserves are lower than anywhere else in the world, restrictions on foreign involvement in many countries and the dependence on national oil companies for a large share of state revenues might constrain the amount of capital available for investment in increasing production capacity.

The principal uncertainties that are clouding future Middle East supply and consequently influencing the region's investment needs are: the rate of growth of global oil demand, the resulting call on OPEC supply and the supply policies that producers opt to pursue. A slightly lower rate of increase in demand than expected would have a particularly marked effect on investment needs in the Middle East as the main residual supplier of oil to the international market. There are also uncertainties on the supply side. The policies of host countries with respect to opening up their oil industries to private and foreign investment, offering access to reserves, the fiscal regime and investment conditions and government revenue needs may constrain capital flows to the sector:

All these factors must be seen against a backdrop of continuing political instability in the region. The US-led occupation of Iraq, endless and exhausting negotiations over a peace deal between Israel and Palestine, and social and political tensions throughout the Gulf compound the political and economic unpredictability of the region. These factors raise the risks and the costs of doing business in the region. Unsurprisingly, country risk is considered most acute in Iraq. It is currently rated lowest in Kuwait, Qatar and the UAE.

Future trends in the natural decline rates of Middle East oilfields will have a major impact on how much investment will be needed. Much of the oil produced in the Middle East today comes from super-giant fields that have been in operation for several decades. For example, the Ghawar field in Saudi Arabia, the world's largest, started production as far back as 1951. The ten largest fields in the Middle East, all of which were discovered before 1975, account for around half of the total output. The natural decline rates at some of these fields are thought to be rising.

The uncertainties described above have implications for the pricing and production policies of the key Middle East oil producers, the role of foreign investors and the timeliness of investment decisions. As long as prices remain high, the governments of the major resource holders in the Middle East will have a very strong economic incentive to ensure that upstream oil investment occurs promptly to extract the very large rents available. However, if for any reason, the necessary amount of investment in the region is not forthcoming and consequently, production does not increase as rapidly as expected, more capital would have to be spent in other more costly regions. To support this additional investment, oil prices would have to increase, leading to a detrimental impact on global economic growth. Oil revenues in Middle East OPEC and other OPEC countries would be lower. These findings imply that it is in the interests of both consumer and producer countries to facilitate capital flows to the Middle East upstream oil sector.

In conclusion, it may be observed that global energy developments have presented the Gulf oil and gas sector with unprecedented opportunities for increasing production levels, expanding future capacity, and funding further economic growth. However, the sector will have to overcome significant investment, infrastructural and policy constraints if it is to capitalize on these opportunities and realize its full potential.

KEYNOTE ADDRESS

The Gulf Oil and Gas Sector: Potential and Constraints

H.E. Abdullah bin Hamad Al-Attiyah

It is well known that the proven oil and gas reserves in the Gulf region are considered the largest in the world. It is certain that with the expected increase in world demand and consumption, the region's future role, if not its responsibility, will be to provide most of the required hydrocarbons to meet the increasing global demand.

Hydrocarbon reserve potential is available in the region, and there is undoubtedly a need for these reserves. What has been achieved in the Gulf oil and gas sector ensures sustained growth of production, industrialization and exports in the region.

The important objective that we must seek to achieve is growth and expansion, not only in terms of quantity, but also in maximum efficiency, so as to guarantee the optimal utilization of our natural resources and of our investments in the processing of these resources.

Constraints may be defined briefly as anything that prevents or hinders the realization of the full benefit from potential. As for potential, we must go beyond its narrow meaning and speak about integrated material and human capabilities. In order to achieve integrated and optimal utilization of our natural resources, we need intensive effort and investment in the development of material, human and social capital.

The development of human resources, including leadership training in all fields, is a primary task if we seek sustained and deep-rooted development.

Technology transfer, for example, cannot be realized in a feasible and optimal way by simply purchasing intellectual property rights and the latest products from others. It also involves having trained national human resources capable of interacting with and using modern technology in a manner that serves our needs and purposes. Another fundamental task is to develop the material infrastructure and basic services, which will considerably reduce the investments needed and shorten the time required for various projects, as well as realize great savings in the overall cost of the development process.

Undoubtedly, at the regional level, there are wide horizons for cooperation. There are many cases in which the fruits of coordination and joint ventures have not reached our expectations. It is necessary to coordinate the plans and fields of industrialization within the region to limit duplication and competition in local and global markets and benefit from the economics of large-scale projects, as well as to make use of contractual experiences and expertise in the best interest of investors in the region, whether government, mixed sector, or even private sector. In a world moving towards globalization, we must develop systems of work and investment that are consistent with specific requirements and the type of business. To do this, we should keep pace with the rules of economic and industrial progress under which we will have to compete sooner or later.

We must benefit from the experiences and expertise of other regional economic blocs and communities, such as the European Union and the European Common Market. Despite having many members, which are mostly developed nations that remain separate entities with their own sovereignty, the European Common Market is seeking to use joint planning in order to improve efficiency and reduce waste in many domains.

There are basic plans for fields such as energy, transport, natural resources, medical affairs and environment. These plans determine the long- and short-term goals. They are general in nature and focus on agreed basic principles and essential objectives, leaving member countries the task of formulating the details of executive plans in a way that suits their own potential, domestic considerations and social characteristics, without prejudice to the common goals and interests of the other members.

The general path is determined by the Secretariat-General, and each country can draw up its own roadmap towards achieving the common goal to be followed up through continuous monitoring and by determining the degree of deviation from the basic track. If such deviation is considerable, it may require a periodical review of plans and objectives, and extending help to some member countries or granting them exemptions so that they can keep pace with the collective progress toward common goals.

The general rule, then, is joint planning, follow-up and cooperation. In this context, cooperation is not used in its material sense. Rather, it means cooperation in identifying the constraints, including the existing hurdles that prevent goals from being achieved. Cooperation may entail certain aspects that are considered by some as linked to sovereignty. Here, there is a need to balance each country's sovereignty-related aspects with the moderating constraints imposed by virtue of collective membership.

It may be useful to mention that sound planning at the country or regional level must be based on complete information, including all data required by the planning process, such as the geological, economic and personal data and industrial assets.

Transparency of information and unlimited availability from original sources are primary requisites that must be given priority over the formalities involved in establishing information centers full of sophisticated electronic equipment and highly qualified professionals working in expensive air-conditioned buildings. What is important is to provide information and statistics, which can be done only by ensuring the appropriate political decision for this purpose and good intentions from all parties without exception.

The time when some information was considered confidential or of strategic value has become history. Globalization, the Information Revolution, and modern technology have made it difficult to stop technology from being shared or communicated. Thus transparency saves a lot of effort and cost, and helps to reduce time and thereby achieve goals.

In conclusion, I would like to stress that the oil and gas sector in the Gulf possesses enormous potential, which can be expanded through cooperation and this sector can prosper by the reduction, if not the removal of existing constraints.

Section 1

PROSPECTS FOR OPEC

1

The Future of OPEC in a Globalized World Energy Market: Opportunities and Constraints

Adnan Shihab-Eldin

In discussing the challenges posed by globalization to oil producers this chapter will focus particularly on identifying and analyzing the opportunities and constraints facing OPEC and examine the role of the Organization in the energy market in the future.[1] It is always instructive at the outset, to define the relevant terms. Therefore, it is appropriate to recall the role of OPEC and then define what is meant by "globalization," before assessing its impact on the world energy market.

OPEC's role has been formally defined by statute, declarations and resolutions, while precedent and pragmatism have also placed their stamp on it from time to time. The most succinct expression of OPEC's role can be found in the OPEC Statute,[2] which dates from 1961, the year after the establishment of the Organization. This contains OPEC's core values: stable markets, reasonable prices, steady revenues, secure supply and fair returns for investors. It reads:

> The Organization shall devise ways and means of ensuring the stabilization of (oil) prices in international markets, with a view to eliminating harmful and unnecessary fluctuations. Due regard shall be given at all times to the interests of the producing nations and to the necessity of securing: a steady income to the producing countries; an efficient, economic and regular supply of petroleum to consuming nations; and a fair return on their capital to those investing in the petroleum industry.

But OPEC's mission does not stop there. Landmark declarations in 1968, 1975[3] and 2000[4] extended OPEC's orbit to cover a broader range of issues.

In 1968, OPEC issued a Declaratory Statement of Petroleum Policy in Member Countries. This referred to the inalienable right, as expressed by the United Nations, of all countries to exercise permanent sovereignty over their natural resources in the interests of their national development. The Declaratory Statement pointed out that this right applied to OPEC Member Countries directly undertaking the exploitation of their own, indigenous exhaustible resources.

In 1975, a Conference of OPEC Sovereigns and Heads of State in Algiers adopted a "Solemn Declaration," which, among other things, embodied a proposal for a New International Economic Order. This aimed at promoting a more equitable global economic system, with particular emphasis on alleviating poverty and other injustices affecting developing countries, by encouraging greater interdependence among nations from the North and the South.

This first OPEC "Summit" led to the establishment of OPEC's own specialist multilateral development finance institution in 1976, intended to help other developing countries pursue social and economic advancement. To date, the OPEC Fund for International Development has made total financial commitments of just over US$7.0 billion. In addition, over the years, OPEC Members have independently provided considerable assistance to other developing countries.

At its Second Summit in Caracas, Venezuela, in September 2000, OPEC's Heads of State and Government reviewed the state of the energy industry, "taking into consideration the rapid pace of change in economic, political, technological and environmental developments, and the challenges and opportunities created by globalization and liberalization." The resulting second Solemn Declaration saw Member Countries reaffirm their commitment to OPEC's longstanding principles and objectives, "to aim at the preservation and the enhancement of the role oil would play in meeting future world energy demand." It also accommodated other important issues which had risen to prominence on the international agenda since the First Summit.

Therefore, OPEC's mission has many dimensions and extends well beyond the confines of the international oil market. Many factors stem from pragmatism, which has helped refine and redefine OPEC's role over the years. With the passage of time, precedents have been established for future actions. In particular, one might refer to OPEC's market-stabilization measures, its unflinching efforts to encourage dialogue and cooperation across the industry, and its readiness to involve itself actively in global debates and negotiations on non-petroleum issues which may eventually have an indirect impact on the fate of the industry, such as climate change and the world trade talks.

OPEC's development has been remarkable, considering its humble origins. It should not be forgotten that, when OPEC was formed in September 1960, five oil-producing, developing countries joined forces to safeguard their legitimate national interests, at a time when the international petroleum industry, outside the former Soviet Union, was under the firm grip of the established industrial powers. At present, 44 years later, as an organization of 11 Member Countries, OPEC is recognized as an established part of the international petroleum community, and its views are sought on all kinds of issues related to the present and future welfare of the industry across the world. Moreover, its production agreements are widely seen as making a major contribution to market stability, to the benefit of producers and consumers alike.

Globalization

Turning the focus of attention to globalization, it may be observed that the petroleum industry has been global in character since its earliest days. Although the industry was born in the United States in the mid-19th century, it grew very rapidly, and the search for, and exploitation of, crude oil reserves soon spread beyond national boundaries and into the world at large.

However, in today's world, "globalization" has taken on a broader and more pervasive meaning and impact. Indeed, on examining the issue more closely, there appears to be no single global definition of globalization! A Google search on the Internet to the posed question: "What is globalization?" yielded nearly two and a half thousand entries! One British newspaper, *The Guardian*,[5] which sought to answer this very question two years ago, began cryptically: "It was the anti-globalization movement that really put globalization on the map!"

An operational definition may be adopted for the purposes of this chapter that describes "globalization" as the process – beyond the full control of individuals and nations – which has brought closer economic, social and political integration among the countries and peoples of the world. Globalization's accelerating strength, scope and wide-scale manifestation has made it a powerful and inescapable force shaping the world. This rapid integration was facilitated and enhanced by the incredible technological transformation that occurred over the last decades, especially in information and communication technologies and transportation. Changes, such as the invention of the Internet and mobile phone technology, have gone a long way to break down borders and barriers that exist between countries to create an interlinked and interdependent global market. Reductions in the cost of communications and transportation have greatly lowered or even erased the barriers hindering the movement of goods and services among countries.

A central theme of the globalization process is the effect it is having on the power of the state and the relationships between states and markets. It is driven by policies advocating the promotion and widespread adoption of market-oriented, competition-based policies, such as privatization, deregulation, the lowering of trade barriers and the opening of capital accounts to international financial flows, as the way to develop greater efficiency and economic development. This process, led by large international corporations, has held out the promise of raising global welfare and bringing prosperity to all regions of the world, through the optimal allocation of resources across the globe.

The results, however, have been somewhat mixed. While it has proved to be a powerful force for restructuring the world economy and, in the process, has brought great benefits to many people across the globe, globalization has also disrupted many other societies and necessitated a number of adjustments in both rich and poor countries to alleviate the impact of this social and economic upheaval. Moreover, globalization has yet to fully deliver on its promise of economic benefits in many developing countries and regions. Indeed, some countries have even seen deteriorating terms of trade for the commodities their economies rely on, as well as an accompanying increase in poverty and unemployment. This has led to disenchantment with the entire process and an increasing backlash against the institutions that are seen to be promoting greater globalization, like the International Monetary Fund, the World Bank and the World Trade Organization (WTO). The lack of positive progress at the WTO talks in Cancun last year is symptomatic of this growing counter-response.

Moreover, an increasingly interdependent and highly complex world, in which nations are linked together by a network of trade and capital flows, is also a more fragile world. Domino effects and spill-over phenomena can quickly magnify any local instability or shortfall into a bigger global issue and lead to a bubble or crash, such as the financial crises in Southeast Asia and Russia in the 1990s. A detailed examination of the growth of certain instabilities and the formation of bubbles and crashes within global financial markets, including energy markets, is beyond the scope of this chapter. However, promising insights into understanding the behavior of financial and commodity markets are being gained from studies that draw on ideas from the highly fascinating topical science of complexity and complex systems.[6]

Globalization has also heightened the need to expand and strengthen the framework of global governance, which, in its current form, is itself weak and inadequate, still evolving and not free from instability. Additionally, global issues, by their very nature, should be tackled at a global level—whether they are security challenges, environment or trade issues, or the promotion of sustainable development.

How do the changes wrought by globalization affect the oil industry? While it is true, as observed earlier, that the petroleum industry has been global since its earliest days, the impact and magnitude of globalization are dramatically different today from what they were a century or more ago.

It is necessary at this point to note the special characteristics of the world energy market: its sheer size vis-à-vis other global industrial and commercial sectors; its pervasiveness and indispensability with regard to its effects on our lives in all but the poorest global communities; and its widely expected expansion in the years ahead, embracing hopefully, an ever-growing spread of the global community, in pursuit of sustainable development.

Today, fuels make up almost a tenth of the world's trade by value with oil representing the major share, and around four-fifths of all oil produced is exported.[7] With this international outlook, the oil industry has already adjusted itself to many changes and grown accustomed to the risks and benefits stemming from a global market.

However, recent developments have presented new challenges to the way the oil industry, and oil producers, in particular, function in a truly globalized world. Realigned political groupings play a significant role in energy globalization, notably the break-up of the former Soviet Union and the expansion of the European Union. Rapid economic growth, of the kind that has occurred in much of Asia over the past two decades, has shifted the centre of gravity with regard to world energy demand. In all such cases, it will be found that in today's world, the global relevance can be as significant as the regional or national relevance.

Major issues that may have a significant direct or indirect impact on the energy market either now or in the future are now discussed or handled extensively at a global level in international conferences or negotiations—on climate change and the Kyoto Protocol, world trade, sustainable development, the nuclear industry and the implications of Chernobyl, and renewable energy.

Moreover, at the corporate level – for everything from day-to-day operations to investment plans for the future – the global dimension is now implicitly understood to be an integral part of any project or task, for many enterprises that do business internationally.

In a nutshell, the oil trade is already highly developed globally.

Short-Term Market Outlook and Challenges[8]

Events in the world energy sector were dominated in 2004 by unstable trends. This has been a major cause for concern among OPEC's Member Countries. Prices for OPEC's Reference Basket of seven categories of crude have recently reached record levels, since this yardstick was introduced in January 1987. They rose above US$40 a barrel for the first time ever in August 2004 and this compares with an average of US$25.8 per barrel from the inception of the OPEC price band in 2000 through 2003 (see Figure 1.1).

Figure 1.1
Frequency Distribution of the OPEC Reference Basket Price
(comparison of empirical with reference distribution)

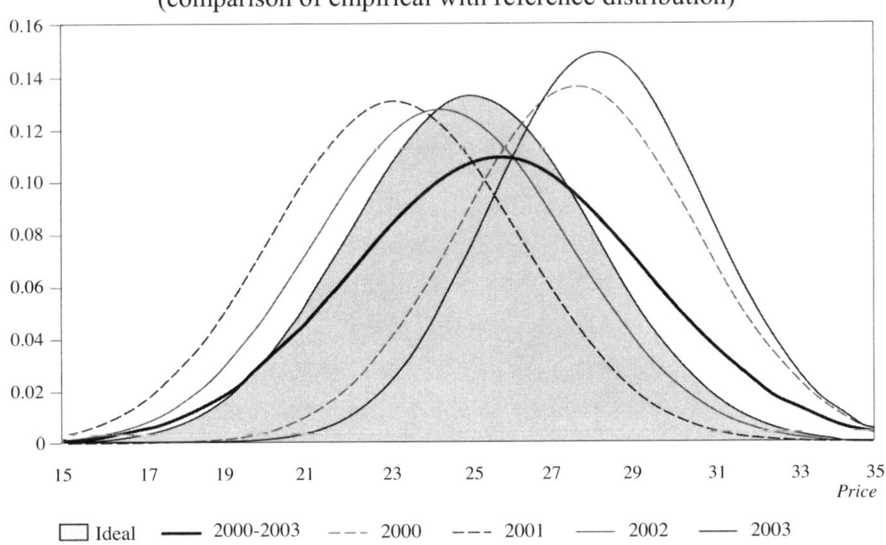

Source: OPEC Data Base

A combination of factors contributed to the rising price trend in 2004, although throughout this period, the market remained well-supplied with crude. The fundamentals have also been sound: higher-than-expected oil demand growth, especially in China and the USA; refining and distribution industry bottlenecks in some major consuming regions, coupled with more stringent product specifications; and the present geopolitical tensions and concern about adequacy of spare capacity to meet possible supply disruptions. In combination, these factors have led to fears about a possible future supply shortage of crude oil, which, in turn, have resulted in increased speculation in the futures markets, with substantial upward pressure on prices.

Indeed, the rapid rise in demand has taken major forecasters by surprise and growth projections for 2004 have more than doubled since this time last year, to 2.5 million barrels per day (mbpd) which represents a 16-year high (see Figure 1.2). This is a 3.2 percent annual growth rate, compared with the 1.6 annual average of 1994–2003.[9]

Figure 1.2
Regional Oil Demand Growth
(year-on-year change)

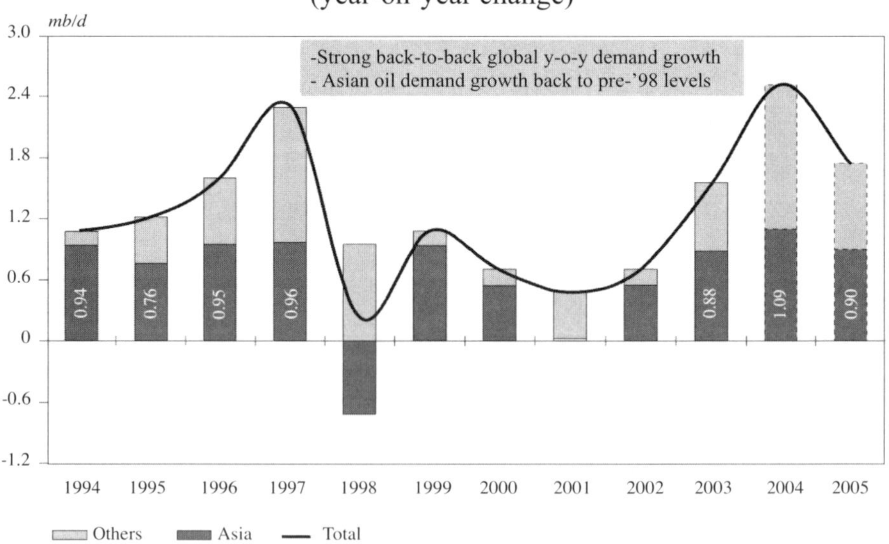

Source: OPEC Data Base

Turning attention to 2005, initial forecasts from recognized sources for 2005 assume a moderate slowdown in global economic growth (see Figure 1.3). We have projected an annual growth rate of 4.3 percent for 2005, compared with the 4.8 percent currently forecast for 2004. Growth will be much faster in developing countries than in the OECD, standing at 5.1 percent compared with 3.0 percent. Separately, China is projected to experience 7.7 percent growth in 2005 and Russia 5.5 percent.

What impact will this have on oil demand in 2005, especially when there may be attempts to fill the strategic petroleum reserves under construction in China and India, as well as, the possible expansion and filling of such reserves in the United States and other OECD countries? Demand growth forecasts for 2005 fall within a wide range of 1.6–2.0 mbpd, with an average of around 1.8 mbpd. OPEC itself projects 1.7 mbpd. Asia is expected to account for a significant proportion of this growth.

Figure 1.3

Growth in World Economy and Oil Demand

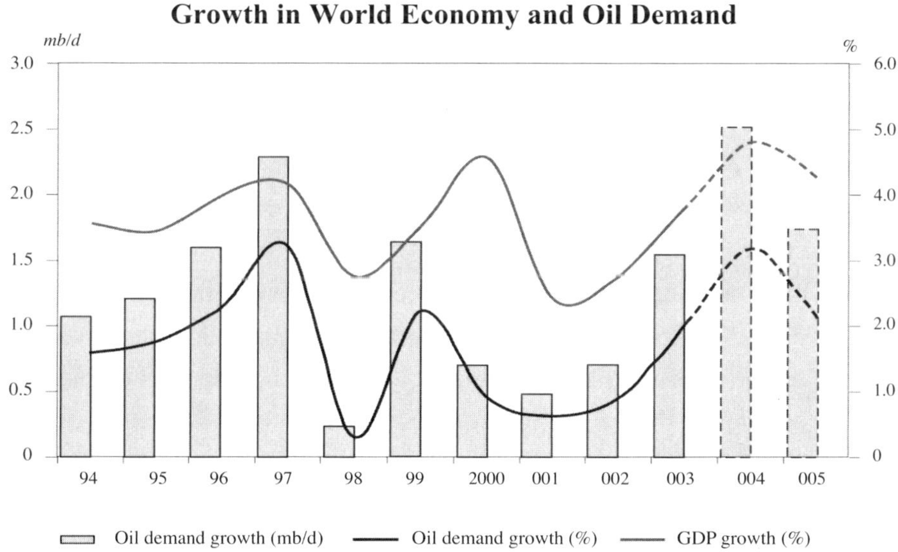

Source: OPEC Data Base

Looking at supply, forecasts for 2005 non-OPEC supply also cover a wide range, between 0.6 mbpd and 1.3 mbpd, with a mean of about 1.1

mbpd. OPEC's projections see an increase of 1.2 mbpd. According to various sources, the difference between world oil demand and non-OPEC supply, representing requirements for OPEC oil, is expected to increase for the third consecutive year. The preliminary market balance forecasts – demand minus non-OPEC supply – are also spread across a wide range, from 27.6 mbpd to 29.4 mbpd, with an average of around 28.6 mbpd. OPEC expects this number, which is, effectively, the call on OPEC oil, to be 28.0 mbpd, at the lower end of the range.[10, 11, 12]

As one would expect, in the light of the present outlook, much concern is being expressed in the world at large about the possible impact of high prices on economic growth. Developing countries, which are the fastest-growing sectors of the world economy, are much more vulnerable than industrialized ones, because of the former's higher oil intensities. The impact on the developing world would be greatest in the large trading economies of Southeast Asia. The emerging economies of Eastern Europe would also be noticeably affected. Moreover, the duration of the high prices is very important, as is the final impact on the consumer—considering the exchange rate effects and the differing levels of fuel taxation. The high level of oil product taxes in European countries, for example, coupled with the recent strengthening of the euro against the US dollar, have effectively shielded consumers in the euro-zone (and Japan) from the rise in oil prices.

Our analysis indicates that the direct contribution of the rise in prices to the economic slowdown in 2004 has been very small. The relatively high level of crude prices (see Figure 1.4) did not appear to have a significant impact on the world economy, because of the strength of the stage of the recovery, as well as decreasing oil intensities in OECD countries. However, a continuation of crude oil prices above US$40 per barrel for an extended period of more than one year could potentially reduce OECD growth by about 0.2 percent, with a more significant impact expected on emerging economies. A recent study by the Asian Development Bank indicated that a sustained 30 percent rise in oil prices from around US$30 per barrel for Dated Brent to US$40 per barrel could reduce GDP growth in Asia by 0.8 percent (without Japan) and by 0.6 percent (with Japan). However, the indirect effect of oil market volatility on consumer and business confidence is harder to judge, especially since these fluctuations are linked to the wider global political uncertainties which could potentially lead prices even higher. Also difficult to judge are the impacts of policy responses.

Figure 1.4
OPEC Reference Basket Price in Nominal and Real Terms* (1994–2004)**
(base: July 1990=100, quarterly data)

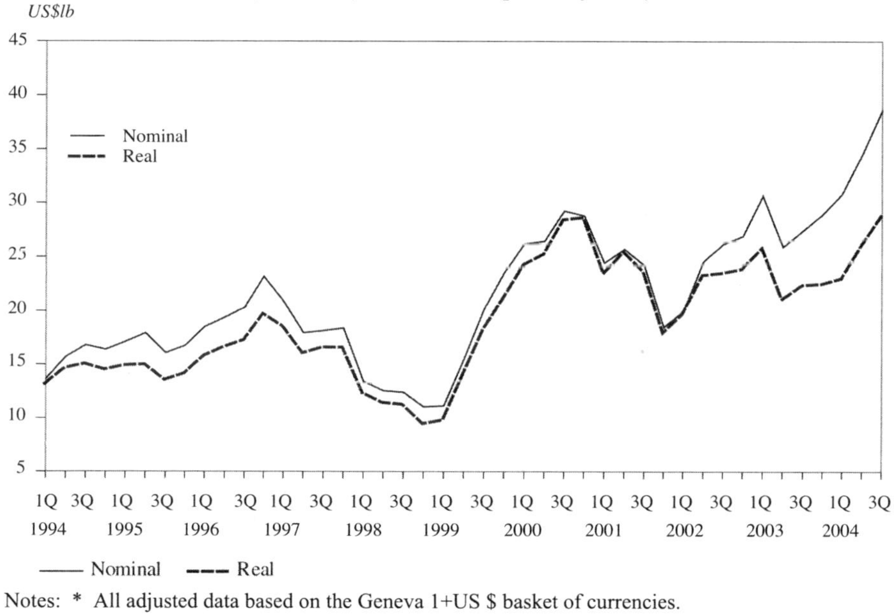

Notes: * All adjusted data based on the Geneva 1+US $ basket of currencies.
** up to 3rd Quarter 2004.

Source: OPEC Data Base

The growth of the world economy in 2005 is expected to be cyclically lower than in 2004. The less favorable cyclical situation may worsen consumer and business confidence, and sustained high oil prices might restrict investment in OECD countries. Developing countries have benefited from the economic recovery in 2004, but the outlook in 2005 is less encouraging, especially considering the uncertain path of the Chinese economy. Lower rates of growth of world trade and high oil prices could have a tangible negative effect, particularly on the emerging economies of Asia and Eastern Europe.

How does OPEC react to all of this? OPEC constantly monitors energy market developments and analyses possible outcomes for the market, at its Vienna-based Secretariat. The Organization is prepared to act at any time – as it has done so often in the past – with the aim of ensuring order and maintaining stability in the market, with reasonable prices that are acceptable to producers and consumers alike.

While the unusually strong demand growth in 2004 surprised market observers, OPEC responded in a timely manner by taking the appropriate remedial measures. At the last two meetings of OPEC's Ministerial Conference – in Beirut on June 3, 2004 and Vienna on September 15, 2004 – Member Countries agreed to raise the Organization's production ceiling by a total of 3.5 mbpd to 27.0 mbpd, with the final increase of 1.0 mbpd coming into effect on November 1. These decisions were taken, even though OPEC's assessments had indicated that there was plenty of crude already in the market and that Member Countries were already pumping out levels of crude well above previous ceilings. It was believed that, apart from the physical fact of agreeing to these big increases in supply to the market, such actions, in themselves, would also send a powerful psychological signal that OPEC was ready to act in order to help reduce prices. Moreover, an Extraordinary Meeting of the Conference was held on December 10, 2004 in Cairo to review market developments and, if necessary, adjust the production ceiling agreement accordingly.

In deciding on production agreements, two considerations have been paramount for OPEC. The first consideration has been to bring prices down to reasonable levels that are acceptable to producers and consumers alike. The second consideration is to keep them there, in other words, to achieve sustainability.

This process, however, has been compounded by the fact that there is a need to establish whether the price volatility seen in 2004 reflects a structural shift in market fundamentals (see Figures 1.5 A+B) or whether it was caused purely by the present exceptional factors affecting the market—or the degree to which it is a mixture of the two. For example, has there been a lasting upward shift in oil demand growth from an historical annual average of 1.7 percent to a level well above two percent a year?

Analysts are now turning attention to another issue that arises when there is continuing demand growth—production capacity. Will this be sufficient to cope with rising requirements in the coming months and the middle of this decade? It is, therefore, useful to spell out the situation affecting OPEC. Historically, the organization has maintained adequate excess production capacity to provide a timely response to supply disruptions and/or surges in demand, such as those experienced in the last few years. With regard to the present situation, it has opened the taps in response to unexpected rises in market demand so far this year. OPEC remains confident that it can continue to do this. To date, it has spare production capacity of around 1.5 mbpd, which would allow for an immediate additional increase in production. Furthermore, in response to the expected demand growth in the future, Member Countries had plans in place to further increase capacity by at least 1 mbpd towards the end of 2004 and through 2005. In addition, plans for additional capacity expansion are available and could be enacted soon. However, this capacity would, typically, become available around 18 months after commencement of this process.

At the present time, although it cannot be said with certainty, the exceptional factors in the market, referred to earlier, may be responsible for adding about US$10–15 to the price of a barrel of crude. Nevertheless, this is a substantial proportion of the price—between around a quarter and a third. In such a situation, there is only limited room for maneuvering with regard to effective market-stabilization measures.

Figure 1.5A
Long Position of Non-Commercials vs. WTI
(weekly futures only, 2000–2004)

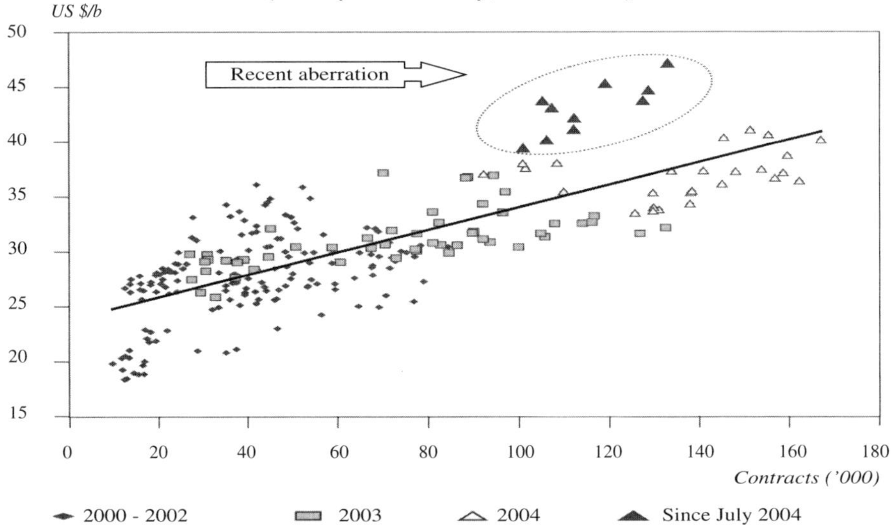

Figure 1.5B
US Commercial Crude Oil Stocks vs. WTI
(weekly, 1994–2004)

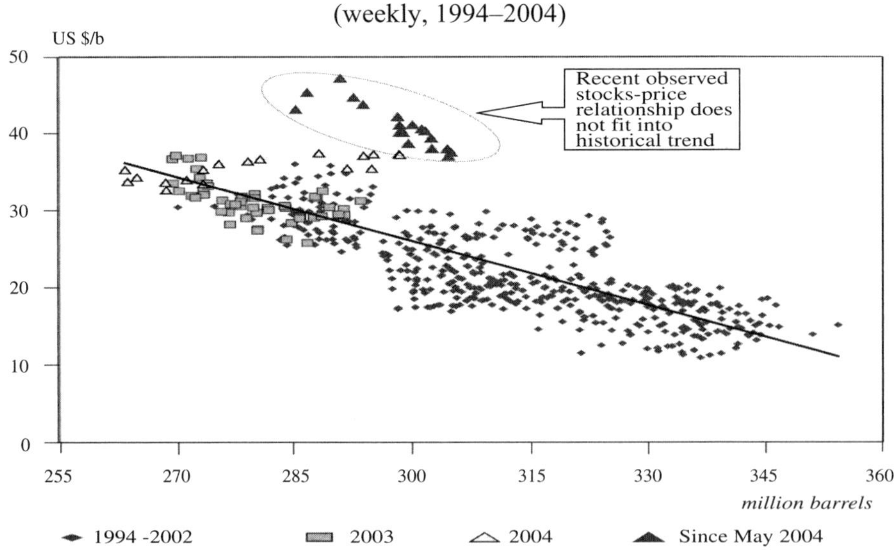

Source: OPEC Data Base; Commodity Futures Trading Commission (CFTC); US Department of Energy (DOE).

Furthermore, OPEC is always aware of another important consideration. What will happen to the oil price if there is a sudden, unexpected dramatic easing of these exceptional factors? If the market is not prepared for this, it could quickly trigger a major price slide, to levels well below those dictated by supply and demand fundamentals—and lower than the US$22–28 per barrel range of OPEC's price band, which won wide acceptance among producers and consumers in 2000–03, before the present exceptional circumstances arose. OPEC must always be alert to this danger and be ready to act accordingly, if such an eventuality arises, or seems likely to arise.

In all these short-to-medium considerations, what is apparent is the concept of uncertainty. As noted earlier, the unusually strong demand growth in 2004 surprised market observers. These included seasoned analysts with the latest high-tech forecasting techniques, hardened industrialists with decades of experience in the industry, and others. Some developments are impossible to predict—although this does not shut out the possibility of employing informed judgment, intuition and astute scenario-building because the industry relies heavily on this. Once again, useful insights into the frequency, timing and magnitude of such uncertain market factors can be gained from the study and application of complexity science in the financial and commodity markets.

Long-Term Market Outlook and Challenges

When looking further and further into the future, the scope for uncertainty increases, perhaps in an exponential manner. The fundamental, interdependent drivers of change in the oil market and the energy market over the medium-to-long term may be pinpointed below:

- The world economy and global issues
- Energy policies and technologies impacting future oil supply and demand
- Oil price paths.

The first two sets of drivers pose the greatest challenge to OPEC and other oil-producing developing countries, since they constitute forces that lie mostly outside producers' direct control. They call for carefully thought-out, deliberate and sustained responses, on the part of producers, so as to anticipate and safeguard themselves against any negative impacts on their interests and sustainable development plans.

Besides the need to cope with uncertain levels of economic growth, a number of key challenges may be identified. These challenges may be noted below:

- Multilateral negotiations and agreements on the environment, trade and sustainable development
- The development of energy technologies
- The energy, environment and commercial policies of consuming countries
- The future of the oil industry and the evolution of the roles of, and the relations between, national oil companies (NOCs) and international oil companies (IOCs).

Given the complexity of the interdependent drivers of future change and the set of challenges they pose, reliance upon scenarios to describe plausible long-term outlooks becomes even stronger. As a result, there are many scenarios around. These scenarios need constant revision and updating, to reflect the ever-changing energy landscape.

OPEC uses its own World Energy Model, "OWEM,"[13] for its longer-term forecasting and this usually covers a period up to around two decades ahead. OPEC has recently produced a revised set of three scenarios to map out the range of plausible futures for a time-frame up to 2020, with an emphasis on the period 2004–15. Each scenario is described according to drivers of change shaping the scenario on the basis of their respective evolution. They provide contrasts for three types of future and have been named respectively: dynamics-as-usual, high growth and low growth. In the three scenarios, considering expected resource use, no scarcity of oil reserves in terms of geological resources available over at least the next two decades is anticipated.

The dynamics-as-usual scenario envisages a future where the drivers of change shaping the scenario are continuing their past patterns and thus no particular departure from current trends is expected. This scenario is based on average economic growth of around 3.7 percent over the period 2004–15, a rate somewhat similar to the one observed in the second half of the 1990s (see Table 1.1). Oil demand will grow on a yearly basis at a robust 1.5 mbpd in a world where global climate change and local pollution will continue to be dominant issues, but without strong enforceable multilateral agreements (see Table 1.2).

Table 1.1
World Average Annual Real GDP Growth Rates per Region (% pa)
Scenario: Dynamics-as-Usual

	2004-2005	*2006-2010*	*2011-2015*	*2016-2020*	*2004-2015*	*2004-2020*
OECD of which:	3.0	2.4	2.3	2.2	2.4	2.4
N. America	3.8	2.7	2.6	2.5	2.8	2.7
W. Europe	2.2	2.1	2.0	2.0	2.1	2.0
OCED Pacific	2.4	2.1	2.0	2.0	2.1	2.1
DCs of which:	6.7	5.1	4.8	4.6	5.2	5.0
China	8.8	6.5	6.1	5.8	6.7	6.5
OPEC	7.1	4.6	3.6	3.4	4.6	4.3
FSU	7.0	3.8	3.3	3.2	4.1	3.9
World (PPP Basis)	**4.6**	**3.5**	**3.4**	**3.4**	**3.7**	**3.6**

Source: OPEC Secretariat.

Table 1.2
World Oil Demand Outlook (mbpd)
Scenario: Dynamics-as-Usual

	2005	*2010*	*2015*	*2020*
OECD	49.8	51.4	52.7	53.8
DCs	29.0	34.4	40.3	46.4
Transition economies	4.8	5.1	5.4	5.7
World	**83.6**	**90.9**	**98.4**	**105.9**

Source: OPEC Secretariat.

Overall non-OPEC supply is expected to continue to increase, rising to a plateau of 56–57 mbpd in the post-2010 period (see Table 1.3), benefiting from continued advances in technology, increased exploration successes and expanded access to new frontier areas. The amount of oil that OPEC will have to produce will gradually increase to around 37 mbpd in 2010 and to 49 mbpd in 2020. Uncertainties over geopolitical developments, and downstream constraints, which initially keep the basket price at an average of US$36 per barrel for 2004, are assumed to ease gradually over the following three years, with the price settling at US$34 per barrel in nominal terms for the rest of this decade, before rising with inflation. In real terms, the long-term price is thereby US$30 per barrel, at today's prices (see Figure 1.6).

Table 1.3
Non-OPEC Supply Outlook (mbpd)
Scenario: Dynamics-as-Usual

	2005	*2010*	*2015*	*2020*
OECD of which:	20.9	20.9	20.2	19.9
USA and Canada	10.6	10.9	10.8	11.0
DCS, excl. OPEC, of which:	16.1	18.2	19.1	19.4
Latin America	4.2	4.9	5.3	5.6
Middle East and Africa	5.6	6.7	7.2	7.2
Russia	9.4	10.1	10.6	10.8
Caspian/Other transition economies	2.3	3.0	3.7	4.0
Processing gains	1.8	2.1	2.4	2.7
Non-OPEC supply	**50.5**	**54.3**	**56.0**	**56.8**

Source: OPEC Secretariat

The high-growth scenario envisages a future where the oil market is characterized by sustained high oil demand, coupled with relatively low non-OPEC supply, resulting in a large call on OPEC oil (see Table 1.4). This future is shaped by high economic growth, allowing developing countries to double their demand between 2003 and 2020, with China increasing by 150 percent in the same period. In this world, it is assumed that geopolitical and economic conditions are growth-inducing, trade negotiations reach a successful outcome, climate change issues are dealt with through non-enforceable multilateral agreements and policies addressing local pollution have little impact on oil consumption, due in particular to technology development. Policy reactions to this strong growth somewhat dampen the increase in demand, but by 2020 global oil consumption has risen to 111 mbpd. Without such demand management responses, even higher oil demand could be envisaged. To highlight the case for high OPEC supply growth, non-OPEC supply is foreseen to grow at a somewhat similar rate to the dynamics-as-usual scenario, to reach a plateau at 55–57 mbpd after 2010. The resulting increase in oil that OPEC will have to supply is thereby considerably higher than in the dynamics-as-usual scenario rising to 54 mbpd by 2020. The oil price is assumed to remain higher, at US$45 per barrel in nominal terms, for this decade, rising with inflation afterwards. In real terms, the oil price settles at US$40 per barrel, at today's prices.

Figure 1.6

Range of Oil Prices in the Historical Context

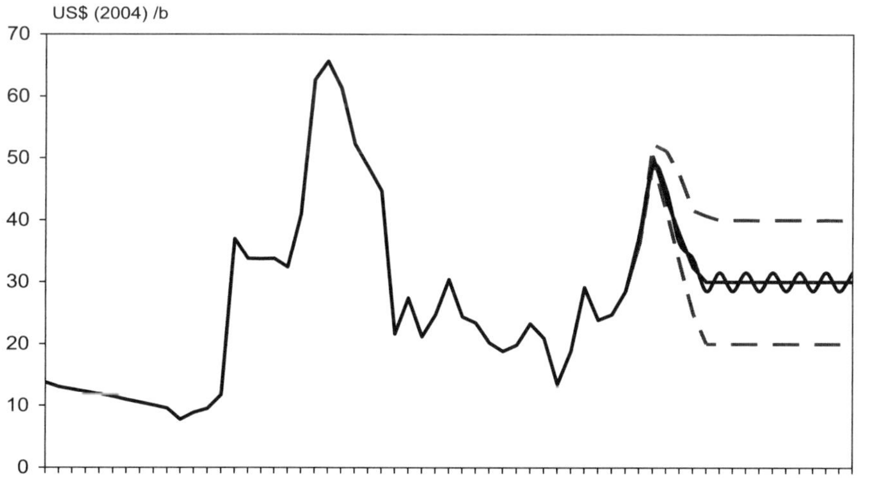

Source: OPEC Secretariat.

Table 1.4
World Supply and Demand Outlook in the Three Scenarios (mbpd)

	2005	*2010*	*2015*	*2020*
Dynamics-as-usual:				
World demand	83.6	90.9	98.4	105.9
Non-OPEC production	50.5	54.3	56.0	56.8
OPEC production	33.1	36.6	42.3	49.1
OPEC market share	39.6	40.2	43.0	46.3
High growth:				
World demand	83.9	93.1	101.9	111.2
Non-OPEC production	50.2	52.7	55.3	57.3
OPEC production	33.7	40.4	46.6	53.9
OPEC market share	40.2	43.4	45.8	48.5
Low growth:				
World demand	83.4	89.2	94.4	99.0
Non-OPEC production	50.8	56.5	57.1	56.4
OPEC production	32.6	32.7	37.4	42.6
OPEC market share	39.1	36.7	39.6	43.0

Source: OPEC Secretariat.

The low-growth scenario depicts a future where the identified drivers of change are compatible with relatively low growth in oil demand and relatively high non-OPEC supply, resulting in low OPEC supply and prices. In this scenario, the world economy will grow at a moderate rate, resulting from increased regionalism, protectionism and tensions among the large trading blocs. Oil demand is strongly managed through enforceable domestic policies and multilateral agreements in developed and major developing countries. Technologies, aimed at using alternative energies in the transportation sector, are increasingly successful and those directed at encouraging oil consumption, such as CO_2 sequestration, are somewhat disappointing. Oil demand reaches just 99 mbpd by 2020, an average annual increase of around 1 mbpd over the period 2004-15. This demonstrates the considerable downside risk associated with future oil demand growth. Non-OPEC supply benefits from a favorable environment, in terms of technologies, levels of new reserves added and fiscal conditions.

It would reach around 57 mbpd in 2010 and plateau at this level afterwards. As a result, OPEC supply will remain low, at 33 mbpd in 2010 and 43 mbpd in 2020. In line with these developments, it is assumed that medium-term oil prices remain at US$23 per barrel in nominal terms, before beginning to rise with inflation at the end of this decade.

The wide variations across these three scenarios underline the profound difficulties facing producers as they seek to meet the challenges facing them in the early 21st century. Three areas are noteworthy.

First, capacity expansion and spare capacity management by OPEC Member Countries will have to accommodate great uncertainty, as the three scenarios result in OPEC supply ranging from 33 mbpd to above 40 mbpd in 2010, and 37 mbpd to 47 mbpd by 2015. To truly appreciate the significance of this, one must consider the long lead times involved in investment in the petroleum sector, which are measured in years rather than months, as well as the importance of "getting it right" by avoiding both excessive over-investment, where the producer is the principal loser, and excessive under-investment, where the consumer is the principal loser. In both cases, the losses and the broader collateral damage, such as to the world economy can be huge.

Second, there is the difficulty of determining optimum price paths for oil for the long term. In real terms, at today's values, oil prices have, with few exceptions, clustered around US$25–30 per barrel over the past two decades with the Reference Basket averaging just over US$25 per barrel in the period 2000 through 2003 (see Figure 1.7). The petroleum industry, however, longs for reliable indications of average future crude prices, to provide a firm base for its investment plans, as well as to add to the general sense of well-being in the market—the "feel-good factor." Moreover, there are the obvious important benefits from reliable oil price forecasting that extend across most areas of world economic activity, particularly where there is heavy international trade and transportation plays a central role in a company's activities.

Third, the downstream sector could be a structural source of price volatility, given the fact that the bulk of the incremental product consumption (from 75 to above 90 percent, depending on the scenario) will occur in developing countries and mostly in China and India.

While OPEC can have a significant degree of influence with regard to addressing the first two of these identified areas of note, the third one – downstream sectors – remains to a large degree, the preserve of consuming nations, even though, in recent years, oil producers have been gaining a bigger presence downstream, too.

In the longer term, the balance of the global oil reserve base and the gradual depletion of non-OPEC reserves mean that OPEC will increasingly be called upon to supply the incremental barrel, with its market share eventually set to rise. By 2020, with a production level – including natural gas liquids – of 49 mbpd, OPEC's share will be above 46 percent of the market.

Figure 1.7
Crude Oil Prices: What do they Suggest?
Nominal vs. Real (1983–2004, monthly WTI US$ per barrel)

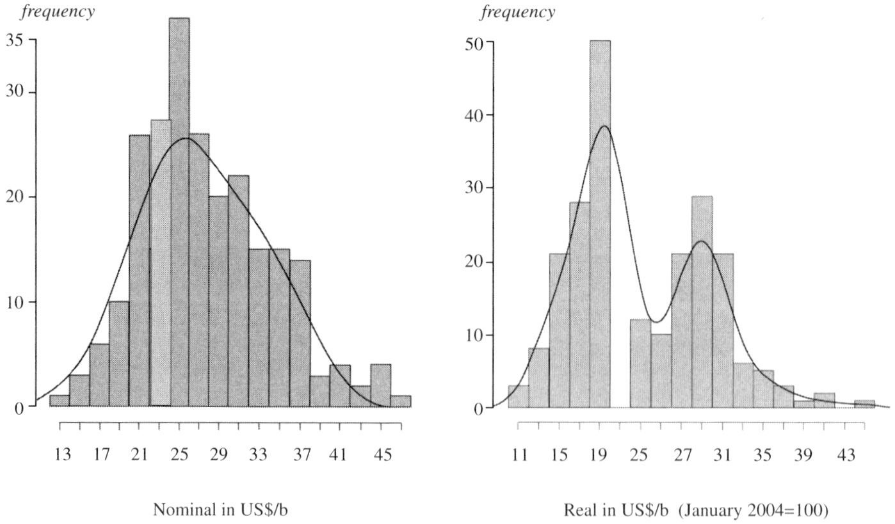

Source: OPEC Data Base.

Clearly there is a need for increased investment in oil production capacity. This has three elements to it. First, it must meet the forecast absolute increase in demand, which was outlined earlier. Second, it must see that exhausted reserves are replaced, as and when necessary. It is estimated that this constitutes the major part of total required investment of around 120–130 billion $ per year, due to the increasing decline rates in most non-OPEC producing fields. Third, it must ensure that oil-producing nations always have sufficient spare capacity available to cope with sudden, unexpected shortages in supply (see Figure 1.8).

Figure 1.8
Estimated Annual Upstream Investment Requirements (2001–2020)

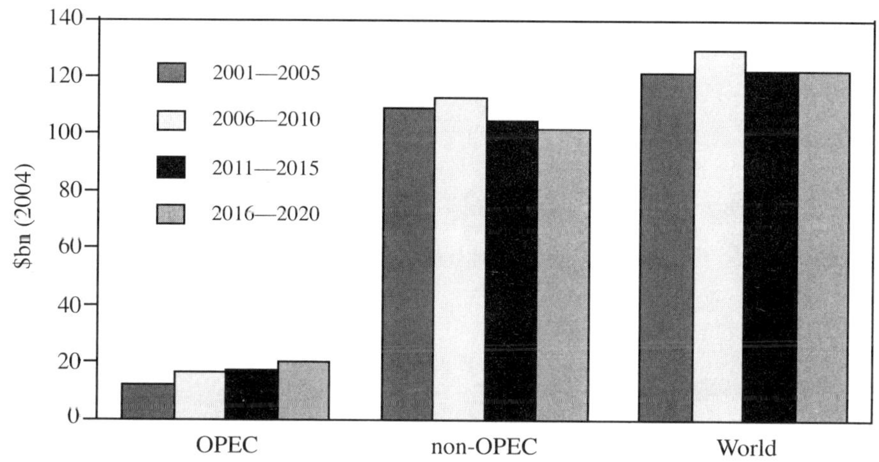

Source: OPEC Secretariat.

However, the magnitude of the required capital injection is far from clear, even in the short and medium terms. This is partly due, as already noted, to the wide range of feasible demand growth scenarios, but it is also reinforced by contrasting views on the potential evolution of non-OPEC production. Uncertainties over future economic growth, government policies and the rate of development and diffusion of newer technologies are among the main factors that lie behind this.

Even a one percent shift in global economic growth expectations can make a big difference to required production capacity levels – and hence investment requirements – in a relatively short time. Our calculations show that a one percent reduction in global economic growth projections reduces the investment requirement for 2010 from a reference case US$130 billion to US$70 billion and, for just five years later, from US$240 billion to US$130 billion. These are very big and indeed, very significant differences. After all, under-investment will lead to a shortage of crude and over-investment will lead to excessive, costly idle capacity. Neither is good, and both can have a serious, detrimental impact on market stability and prices (see Figure 1.9).

Moreover, when a miscalculation occurs, it is by no means an overnight task correcting the situation, due to the long lead-times involved in investment in the petroleum sector. It could take years to make the appropriate adjustments to production capacity levels which are found to be out of line with the market's requirements.

Countering the challenges, therefore, becomes a very important task and to be truly effective, requires a collective approach from within the industry, since the challenges are too large, too complex and too important to be left to partial interests. This requires transparency, consultation, meticulous planning and careful scheduling by all parties.

One certainty, however, is the clear economic advantage that lies in investing in the exploitation of OPEC's crude oil reserves. Not only are these reserves more abundant, accounting for around four-fifths of the world's proven total, but they are also more accessible and cheaper to exploit than those in non-OPEC areas.[14] Also, as mentioned earlier, our projections show that OPEC will increasingly be called upon to supply the incremental barrel. Moreover, it is worth noting that the absolute size of investment required by the oil industry in this outlook is not necessarily different in magnitude to that observed in the past. All this should be viewed as good news for the industry and for world energy supply generally.

Figure 1.9
Cumulative OPEC Investment Requirements

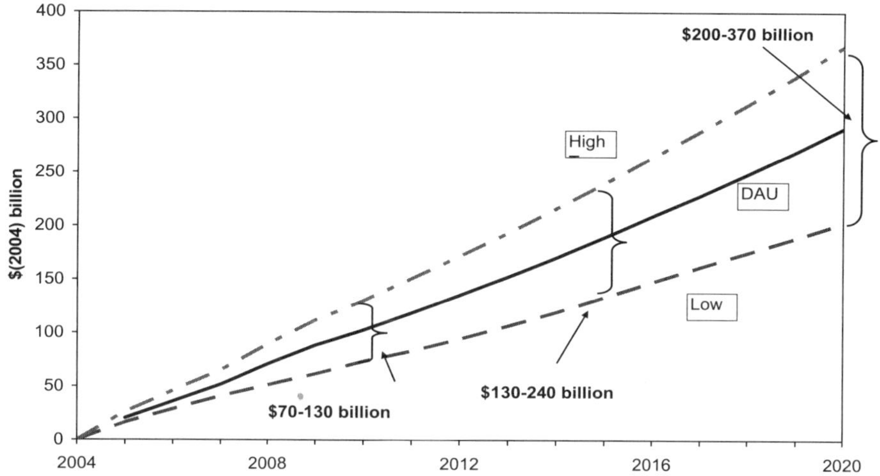

Source: OPEC Secretariat

However, the conditions have to be right to ensure that this oil is brought to the market in an efficient, effective and timely manner. One of the clear messages to emerge from the Second Joint Workshop on Oil Investment Prospects,[15] held by OPEC and the International Energy Agency (IEA) in April 2004, was that the starting point for a sound investment strategy is market order and stability today. A reasonable, predictable price level is needed to secure adequate sources of investment. That is why OPEC's market-stabilization measures and, in particular, its price band concept are so important.

This underlines the importance of cooperation within the industry and it is heartening to note that major advances have been made in this area over the past two decades. All producers stand to benefit from cooperation, on all time-horizons. There must be reasonable burden-sharing among all parties in the industry, including the large oil companies, financial institutions and other intermediary bodies, if the industry is going to develop in an orderly manner in the future and meet the sizeable challenges facing it.

There have also been major advances in producer-consumer dialogue in the 1990s, in tandem with the rising profile of the International Energy Forum. OPEC has played a prominent role in the development of this specialist producer-consumer forum; among other things, its new permanent Secretariat has been located in an OPEC Member Country—Saudi Arabia.

Recent years have also witnessed the development of a closer working relationship between OPEC and the IEA, to exchange ideas and information. Two Joint Workshops on Oil Investment Prospects have been the outcome, while informal discussions between the two organizations in 2003 helped to stabilize the oil market at the time of the Iraq war.

Clearly, the concept of cooperation within the world oil industry is now well-established—although such is the complexity of this industry and the underlying forces and pressures that propel it, that there remains plenty of scope for improvement. Nevertheless, a clear realization has emerged that the industry is better off if there is an underlying consensus on the means of handling at least the major issues that concern all parties—such as price stability, security of demand and supply, investment, environmental issues and sustainable development.

Beyond the Scenarios: Pressure Points Identified

The scenarios presented in this report have portrayed a set of contrasting possible future situations. The driving forces behind the momentum for oil demand growth, as well as that for non-OPEC production could feasibly combine to produce a considerable range for the amount of oil expected to be supplied by OPEC over the medium-to-long term (see Figure 1.10). In line with these developments, it has also been shown that a wide range of oil prices could be associated with these overall scenarios. The price assumptions for each scenario can also themselves be seen as central values of ranges, each of which reflects a qualitative view of the pressures upon the crude oil price that might evolve from oil market fundamentals.

Figure 1.10
A Wide Range for Required Future OPEC Oil Production

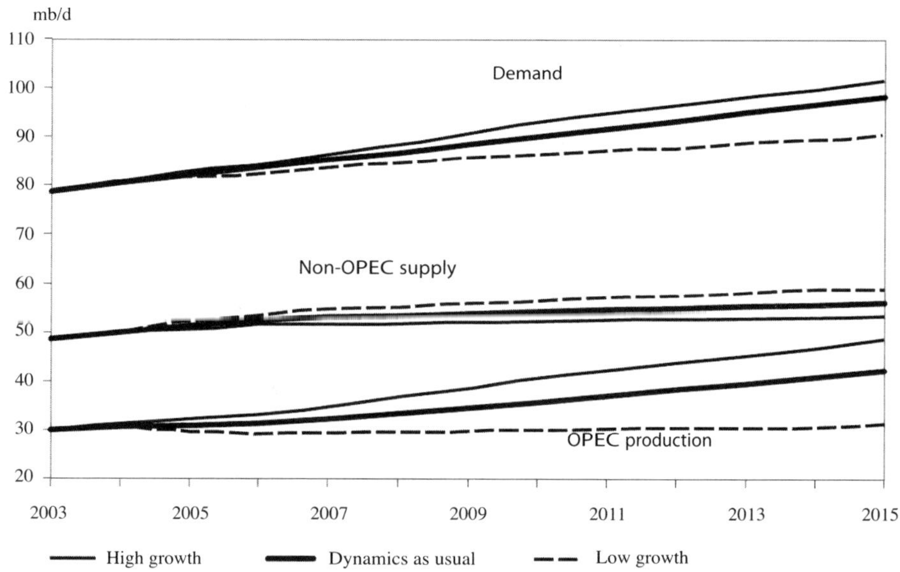

Source: OPEC Secretariat.

Significant pressure points can be identified in the development of these scenarios, and they lead to a series of questions that demand attention. In the dynamics-as-usual scenario, the following issues need to be addressed:

- How can spare capacity best be managed? Can this responsibility be shared in new ways?
- What can be done to tackle the impact of the downstream on oil price volatility?
- Can the effects of speculation on the oil price be reduced?
- Is the price assumed in this scenario at an "appropriate" level?
- What is the best way to ensure that the appropriate investment takes place, bearing in mind that it will be concentrated increasingly in OPEC Member Countries?
- What role will the international oil companies play and how will this affect the ability of OPEC to manage its collective oil production?

In the high-growth scenario, characterized by protracted market tightness, the pressures of the short term in the dynamics-as-usual scenario are expected to continue into the future: strong demand and low non-OPEC levels combine to keep prices at $45 per barrel for the rest of this decade. OPEC clearly will have difficulty in keeping up with the increased demand for its oil. The sustainability of such an evolution may be questioned.

To begin with, this market tightness, if, indeed, prolonged, could use up almost all spare capacity, leading to a spike in prices, rather than the smooth path that is assumed.

In the longer term, the high prices could give rise to an eventual reduction in demand growth rates, as the impacts upon GDP growth are felt, while policy and technological developments also respond. A rise in non-OPEC supply could also be envisaged in such a high oil price environment, so that substantial increases in OPEC spare capacity could emerge. This, in turn, could suggest that a fall in prices would become increasingly inevitable. Several pressure points leading to the bifurcation of the future market outlook are suggested by this scenario:

- The drivers causing this tightness could, if prolonged, eliminate all spare capacity and lead to dramatic spikes in prices. What are the best ways to cope with this?
- If the high prices give rise to medium-term increases in OPEC spare capacity, a fall in prices becomes increasingly inevitable. This raises again the issue of a sustainable price path.
- The high growth in the required level of OPEC oil production implies considerably higher investment than in dynamics-as-usual. Even with higher oil prices, to what extent do the NOCs of OPEC countries possess the financial and operational capabilities to supply larger volumes?
- The downstream sector, and in particular the refining industry, could add significant volatility to oil prices due to its unprecedented expected growth. How should this issue be handled?

Finally, in the low-growth scenario, which is characterized by a prolonged soft market, weakening economic growth and a strong wave of policy measures ratchet down the required level of OPEC oil production and the oil price. Consequently, oil prices remain at US$23 per barrel in the medium term, before beginning to rise at the end of this decade with inflation. Prices could soften further, if policy reactions lead, for example, to improved fiscal conditions in non-OPEC countries, in the face of such oil prices. Another potential snowballing effect could relate to the reaction of OPEC's Member Countries to this low price environment: if efforts are made to compensate low revenues with additional output levels, then even lower oil prices could emerge over the medium-to-long term. Alternatively, the low price (and especially the even lower prices that could emerge from the snowballing) may give rise to a medium-term strengthening of demand and a slowing of non-OPEC output, suggesting that a rebound of prices would become increasingly likely. This prolonged soft market, therefore, also contains the seeds of potential market instability, with the longer-term outlook characterized by the possibility of substantial cyclical swings, under pressure from market fundamentals.

Technology and the Future of Oil

On the supply side, revolutionary upstream technological changes have shortened the time it takes to find and develop an oil field, and lowered the cost. With continued advances, non-OPEC output may continue to increase up to 2015 and peak around 55 mbpd, or it could be above 57 mbpd by 2020, depending on the pace of technology development. In OPEC Member Countries, the application of these technologies has the potential to reduce upstream costs further and increase recovery rates, and may, in addition, open up frontier areas. Nevertheless, tapping this potential, to increase the supply of oil and natural gas from these OPEC resources, requires more focus on issues of technology transfer to producing countries, where the use of many of these advanced technologies has not yet fully materialized.

In particular, focus is needed on the development and use of more advanced oil and gas production technologies that are suited to the special characteristics of present and future large oil fields, typically found in some OPEC countries, where most oil reserves and resources are located. Future technological progress would allow the development of significant amounts of unconventional oil at lower cost and may enable the extension of the availability of oil supply in the long run. In addition, gas-to-liquids (GTL) technology is making increasing amounts of clean fuels available.

Technology development on the demand side has focused on reducing demand. Over the past 20 years, the bulk of the demand growth has come from the transportation sector, particularly in the OECD countries. This trend is expected to continue for the next two to three decades, accounting for about 60 percent of the growth. As noted earlier, there are sufficient conventional and unconventional oil resources to meet this projected increase. Nevertheless, to meet the key energy policy objectives of OECD countries in reducing future oil imports and cutting back on urban pollution and harmful emissions, as well as preparing for the long-term eventuality of fuel-switching away from hydrocarbons, efforts are already under way to develop cleaner alternative fuels, coupled with more efficient engines and systems. Improvements in the efficiency of conventional internal combustion engines could achieve a fuel economy of about 40 miles per gallon for the average passenger car within the 2010–15 timeframe. Hybrid vehicles, which can achieve up to 60 miles per gallon (mpg), are available now, and may witness strong growth within the next five years, as performance, economics and safety continue to improve. In the very long term, the hydrogen fuel cell probably represents the vehicle technology which poses the greatest threat to oil's dominance in the transportation sector. However, fuel cell vehicles are currently very expensive, and it is likely that, even if they are produced in large volumes, they would continue to cost more than conventional vehicles. Besides, hydrogen fuel, which is an intermediate energy carrier and not an energy source, is likely to be produced from hydrocarbon sources.

Figure 1.11
OPEC Production in Efficiency Scenarios

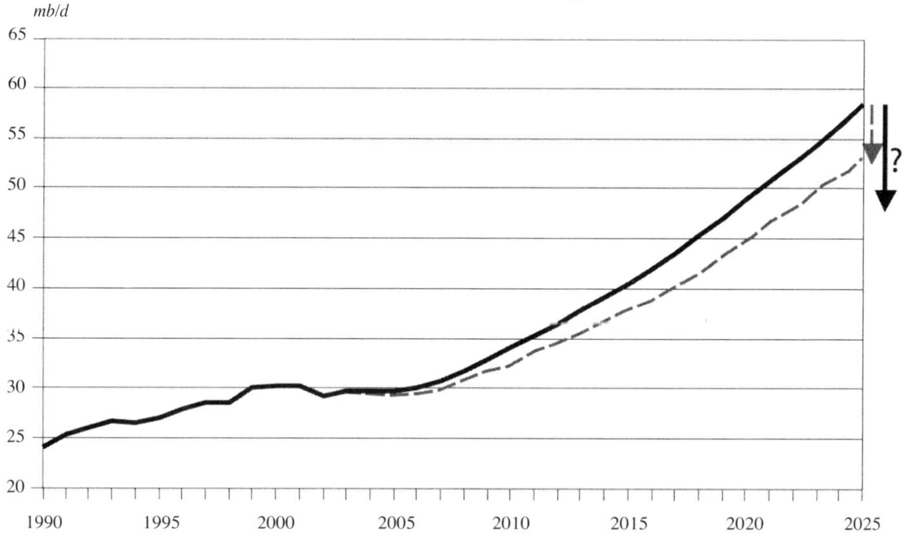

Source: OPEC Secretariat

A scenario that assumes faster advances in average vehicle efficiencies, projects that oil demand will grow to less then 96 mbpd by 2015, a reduction of 2.7 mbpd from the reference case. By 2025, the reduction in demand would exceed 5 mbpd. A more aggressive scenario projects that oil demand in the United States alone could be reduced by up to 2 mbpd, if the average fuel efficiency of the fleet increases by 10 mpg (see Figure 1.11).

Technology could also be developed to sustain and enhance demand for oil. In the face of the perceived climate change challenge and the resulting current calls to effect deep reductions in emissions from the levels shown in the dynamics-as-usual scenario, the use of carbon dioxide (CO_2) capture and storage technology is a promising mitigation option, that will allow the continued use of fossil energy resources in a carbon-constrained world.

Storage of CO_2 in depleting oil reservoirs has been successfully demonstrated and has the potential to increase recoverable reserves and extend the global plateau production rate. If anthropogenic sources of CO_2 are used, there would be a win–win scenario, in which a valuable resource – oil – is made more plentiful by reducing harmful CO_2 emissions.

Carbon dioxide capture and storage (CCS) technologies, applied in OPEC oil fields, could be used to store large quantities of CO_2, enhancing oil recovery, while substantially reducing CO_2 emissions, thus creating space for the continued use of oil in the transport sector in a greenhouse gas-constrained world. The commercialization of integrated oil gasification combined cycle power generating technology is another example where oil may become competitive with gas as a clean fuel for power generation, especially if gas prices increase, thus reversing the trend of declining oil use in this sector.

OPEC and other producing countries stand to benefit by promoting, encouraging and participating in the development and rapid deployment of CCS technology. Considering the high level of R&D funding for CCS in many OECD countries (USA, Japan and the European Union), aimed at realizing commercial zero emission coal power generation, it is highly likely that the technology will be developed with or without OPEC. Participation would give OPEC a "seat at the table" and the possibility to influence policies and development.

There are a number of potential avenues for OPEC and other producers to choose from, ranging from joining existing initiatives and programs, like the IEA Greenhouse Gas Program or the CO_2 Capture Project, to formulating and launching their own international collaborative program geared specifically to CO_2 sequestration and enhanced oil recovery in depleting fields.

Summary and Conclusions

OPEC's philosophy and objectives have evolved over the past four and a half decades, in keeping with the changing conditions of the world energy sector—particularly in oil. Throughout, the organization has maintained its commitment to its core values of stable markets, reasonable prices, steady revenues, secure supply and fair returns for investors. Over the years, OPEC's sights have broadened to cover issues pertaining to the general welfare of mankind, such as sustainable development, the eradication of poverty and environmental harmony.

In depicting the setting of a globalized world energy market, it is clear that the oil trade had already become highly developed globally. An examination of the short-term oil market outlook and challenges reveals some interesting insights. These include the centrality of oil to mankind's energy needs, its dynamic and constantly expanding nature, its inherent volatility and its efforts to cope with the present and cater for the future. OPEC's important role in seeking to achieve order and stability remains very much in focus, at a time when prices were unusually high and yet fundamentals were sound and the market was well-supplied with crude.

OPEC faces obvious difficulties in meeting such a challenge, and these revolve around the organization's commitment to order and stability, fairness, consistency and the reduction of uncertainties. Above all, the need for a collective approach to handling the challenges confronting an increasingly globalized market was stressed.

What is required is a sound vision for the future. Today's challenges are essentially the same as tomorrow's challenges, as indicated in the section on the long-term market outlook. However in the coming years and decades, these challenges will be even larger. Demand will continue its path of relentless growth. Peering deeper into the horizon, the uncertainties are likely to increase exponentially. The world would demand increasingly cleaner and greener fuels and the rightful interplay of energy and sustainable development would gather pace. Technology would achieve further breakthroughs and, no doubt, add one or two total surprises!

However, the same fundamental issues would remain in place: order and stability; secure supply; reasonable prices that are acceptable to producers and consumers alike; and fair returns to investors. These form part and parcel of OPEC's self-proclaimed legacy, a legacy which had been developed since the middle of the 20th century and which had steered the organization and its 11 Member Countries into the direction of effective, responsible market-stabilization measures, from which the whole market has benefited.

The issue is how to meet these challenges. Three contrasting scenarios consider the period up to 2020: dynamics-as-usual, high growth and low growth. Wide variations across these three scenarios underlined the profound difficulties that producers are likely to encounter as they seek to meet the challenges confronting them in the early 21st century. In particular: accommodating great uncertainty in planning the capacity expansion and spare capacity management of OPEC Member Countries; the difficulty of determining optimum price paths for oil in the long term; and the possibility that the downstream sector could be a structural source of price volatility.

In the longer term, the balance of the global oil reserve base and the gradual depletion of non-OPEC reserves would mean that OPEC will increasingly be called upon to supply the incremental barrel, with its market share eventually set to rise.

A sound investment strategy for the future is necessary and the sums to be raised will be huge. A return to the core issues is required—the provision of adequate production capacity at all times, the reduction of uncertainties and collective responsibility. Transparency, consultation, meticulous planning and careful scheduling by all parties are also vital ingredients. These are all values that OPEC has espoused and sought to act upon, whenever and wherever necessary.

In conclusion, therefore, the picture is one of a globalized world energy market with potentially sound prospects for oil and, in particular, OPEC oil in the opening decades of the 21st century.

OPEC can take comfort from the opportunities that lie ahead. Due to its reserve strength in gas and oil there is a powerful potential for its Member Countries to enhance their role as energy suppliers and thus continue to earn revenue, which will enable them to improve the economic and social welfare of their populations. The very long term benefits for sustainable development are obvious here, both for these countries themselves, as well as for other developing states that are benefiting from the assured availability of flexible, cleaner fuels that are conducive to robust economic growth.

This positive outlook is helped by the fact that the industry is more sophisticated, better organized, less confrontational and more unified than ever before, in its desire to ensure steady, secure supply in the future. The increasingly globalized nature of the petroleum industry can only be for the good, by increasing the size of the potential market and by rationalizing the process of supply and demand at world level.

However, some major constraints remain. Many of these are evident today—excessive speculation, volatile geopolitics and unjustified and damaging fears about supply security. There are also constraints in a more physical sense. In particular, reference must be made to investment—the problems of uncertainty, of knowing precisely how much capital to commit to increasing production capacity, in order to ensure that future demand will be met in a full and timely manner.

There is also the constraint of complacency, which affects all industries at some time or another. When basking in the warmth of the sun for a few weeks, it is all-too-easy to forget that storms can brew overnight or that there are three other seasons of the year when the conditions are less friendly!

And then there is the technology factor. A major potential constraint is lack of access to the very latest technology. Technology can contribute to demand growth, just as it can reduce demand. Producers must redouble their efforts and join international collaborative technology development that advances sustainable demand on clean oil. An option that offers great potential involves carbon sequestration technologies, combined with enhanced oil recovery.

In an industry where margins are critical and where use of the very latest technology can yield dramatic returns in an oilfield's output, it does not make sense to global supply as a whole if state-of-the-art technology is denied to a particular development. If a field, during its natural lifetime, has been performing at five per cent below its potential, due to restricted technology, it is not possible to return to it at a later date to retrieve that five per cent. It is gone for ever. Not only would the producer be the loser, but so would world supply.

In the final analysis, the world's hydrocarbon resources are finite, however much recovery rates have improved in recent years and will continue to rise in the future. Therefore, OPEC's underlying role and its contribution to market stability are needed now more than ever, given the increasing globalization and complexity of the oil market and energy markets in general.

It is increasingly being recognized that while in the long term markets should be expected to behave efficiently, as stipulated by the neo-classical economic paradigm, volatility is inevitable in the short and medium terms. Large degrees of instability may arise unexpectedly at any time, with the potential to grow, if left unchecked, with devastating effects.

All markets need some form of prudent regulation and management. Markets should be free enough to function efficiently, but some form of mechanism should be available and close at hand to counter excessive volatility and harmful trends, as and when the need arises. The oil market is far from unique, in this sense. It has been noted recently, for example, that OPEC's excess capacity, and the liquidity it provides to supply, is as powerful a tool for oil market stability as money supply and interest rates are for sustained growth in advanced economies and the world at large.

All considered therefore, there are both wide ranging opportunities and constraints for OPEC producers in the future and the overall challenge facing these countries is to effectively maximize the opportunities and to minimize the constraints.

Oil Market Trends: A Post-Script

Since the original conference presentation made in late 2004, the market trends identified at that time have continued to develop, with nominal crude oil prices reaching new highs, even though they are still below the real levels achieved a quarter of a century ago. This post-script section is intended to update the market situation as it stands at the beginning of November 2005.

The price of the OPEC Reference Basket exceeded US$60 a barrel for the first time briefly at the end of August 2005, although since then it has remained below that level. As of June 16, 2005, a reconstituted Reference Basket has been used based on the average price of a barrel of 11 crudes, instead of seven, as had been the case since this yardstick was introduced on January 1, 1987.

Figure 1.12
OPEC Reference Basket price in Nominal and Real Terms: 1980 to October 2005
(Base: October 2005=100, monthly data)

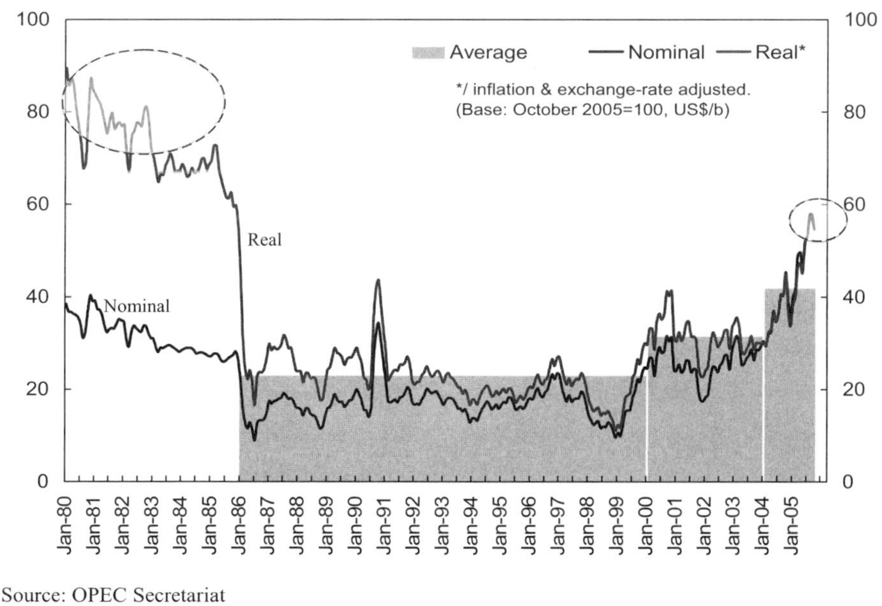

Source: OPEC Secretariat

Looking at the current market situation, it may be noted that a combination of factors continues to be at work, leading to persistent price rises and volatility. There has also been a further shift in the focus from the upstream to the downstream.

The spotlight was initially on the upstream sector, with the unexpectedly high oil demand growth in 2003 and its acceleration to exceptional levels in 2004. This was connected, in particular, to the high level of economic growth of China and the United States. It was exacerbated by a notable development on the supply side, when the growth in non-OPEC supply started to fall behind that of world oil demand, reversing earlier trends of matching or exceeding demand growth. However, during 2004, it became steadily apparent that the destabilizing forces lay increasingly in the downstream rather than in the upstream sector, and this trend has continued to the present day.

Indeed, in 2005, while demand growth has remained at healthy, though more modest and sustainable levels, the challenge of providing adequate crude supply is being met successfully. This has been largely due to OPEC's actions, raising its production by more than 4.5 mbpd since 2003, as part of its market stabilization measures. This in turn, has led to a steady rise in OECD commercial oil stocks, which are now exceeding their five-year average.

OPEC has also been acting on a second front. Assuming that the high levels of oil demand growth – of about 1.5 mbpd per year – are likely to remain at least in the medium term, OPEC Member Countries have sought to accelerate their plans to bring new production capacity on-stream to meet continued demand growth and to re-establish a comfortable level of capacity and help calm markets during the present troubled period. This spare capacity, which is now at 2.0 mbpd, will be more than adequate to cover oil demand growth throughout this winter and in 2006, with the call on OPEC being around the 2005 level. More increases in capacity have been planned and are being implemented for the rest of the decade.

The year 2006 is expected to see non-OPEC oil supply growth recover to 1.6 mbpd from the exceptionally low level of 0.7 mbpd witnessed this year and this figure includes non-OPEC natural gas liquids and non-conventional oils. On balance, the expectation is annual average growth of around 1 mbpd in non-OPEC supply up to 2010, slowing thereafter. Combined with the planned increase in OPEC capacity from 32.5 mbpd to more than 38 mbpd by 2010 and an additional 1.5 mbpd increase in OPEC natural gas liquids over the same period, this means that cumulative world oil production capacity will rise by around 12 mbpd or more, over the next five years. This will be well above the expected cumulative rise in demand of 7–8 mbpd over the same period, and thus it will more than cover the forecast growth in demand.

Moreover, the fact that the market has, over the past two years, experienced and successfully handled a serious spare capacity limitation, together with the fact that the greater revenues generated by the recent higher oil prices have allowed more funds for reinvestment in the industry – compared with the under-investment in new capacity that occurred previously at much lower prices – is a very positive sign for the assurance of crude oil sufficiency well beyond the first decade of this century.

However, there is a different picture downstream. The serious downstream bottlenecks continuing in some major consuming countries – due mainly to a lack of timely investment and to increasingly stringent product specifications motivated by environmental concerns – have seen refineries operating at near capacity to keep pace with rising demand. Not only is this putting more pressure on product prices, but its effects are also felt subsequently on crude prices, especially light, sweet blends. Clearly, the industry at large must pay more attention to the downstream part of the supply chain, in the interests of overall market stability. As it stands now, it does not appear that the current planned growth in refinery capacity will match demand growth before 2007.

On top of all this, there has been widespread concern about possible future supply disruptions that may result from increased geopolitical tensions or other causes. The cumulative effect of all these factors has been reflected in increased speculation in futures markets, particularly through a rise in activity by non-commercials, notably pension and trust funds, which has led to a rise in open commitments. Both types of funds have correlated strongly with the price increases of the last two years. Without any doubt, due largely to the perceived capacity constraints, the market has become very nervous and over-responsive to external impulses.

2

Maintaining Cohesion within OPEC: Implications for Cooperation among Member Countries

Luis E. Giusti

The birth of the Organization of Petroleum Exporting Countries (OPEC) was a direct consequence of the poor vision exhibited by the leaders of the international oil companies. The agreement reached in Baghdad on September 1960 by Iraq, Iran, Kuwait, Saudi Arabia and Venezuela, with Qatar as an observer, was a direct reaction to the unilateral lowering of oil prices by the international oil companies. In early 1959 BP made a unilateral cut in its oil prices, which triggered a torrent of complaints and denunciations from the oil exporters. As Daniel Yergin describes it in *The Prize*, "with the stroke of a pen, a major oil company had unilaterally slashed the national revenues of the oil producers. The exporters were galvanized into action." Clearly, the lesson was not properly learnt, because on August 9, 1960, with no direct warning to the exporters, the Standard Oil Company of New Jersey cut its posted prices of Middle Eastern crudes significantly, with a reduction of 7%. The other oil companies followed suit although without any enthusiasm. Their decision was to prove a fatal move.

These were the circumstances that led directly to the establishment of OPEC by the oil producing nations, with the primary objective of defending the price of oil. The founding members also called for a system regulating production, committed themselves to solidarity in case the oil companies sought to impose sanctions on any one of them, and vowed to ensure stabilization of prices, eliminating harmful fluctuations. At that time the founding group was the source of 80% of the crude oil exports—a strong reason for companies to apologize and reverse their decision. From then on, the companies began to be cautious about making any unilateral moves. It was in fact a turning point, provoked by the oil exporting nations with the objective of controlling their natural resources. For decades those countries had been selling their oil in the international market, but received a relatively minor economic return, despite the wide margin of rent that is implicit in the commodity.

Thus the creation of OPEC was a very important first step taken by the producing countries, but difficult times lay ahead, because the oil companies still retained control over market access. It was not until late 1973, that OPEC assumed control over the fixing of oil prices, independently of the oil companies. Another important evolution was the process of nationalization in most OPEC countries, which transformed the oil companies into mere lifters of OPEC crude (a point that will be discussed in more detail). Paradoxically, those events translated into market volatility, because many OPEC decisions became increasingly influenced by external political events. During the period 1973–1979 major disruptions occurred in the oil market, with prices escalating ten-fold. However, after the supply balance was restored, the appetite for high prices had settled in within OPEC. The high price brought radical changes to the markets in terms of energy conservation, energy efficiency, reduced demand for OPEC oil and an upsurge of new oil resulting from investments in non-OPEC production. Along the way, OPEC lost market share, and when prices began to weaken, the organization decided in March 1983 to engage in a system of supply management, implementing a mechanism of ceiling and quotas. With that step, OPEC assumed the *de facto* role of supplier of the last resort, also assuming, though unofficially, the responsibility for balancing supply and demand. But the system, as explained by Dr. Fadhil Chalabi, is inherently flawed, because in contrast to the oil companies in the past,

OPEC has no access to the latest data about stock movements and stock build-up at the levels desired by refiners, market requirements of products and seasonal adjustments, as such information is available within the oil companies. So in essence, the OPEC's supply management tends to be deficient, and thus a source of price volatility. However, after four and a half decades it can be said without hesitation, that OPEC has had reasonable success. However, the vision, the direction and the cohesion of the organization as a group remain weak. To establish a common cohesive interest, as it was the case in 1960, is easier said than done. Time has brought divergences among the various members to the surface. The particular needs and concerns of each member state are frequently difficult to synchronize into effective unified action. In assessing the future of the organization, there are important reflections to be made and important questions to be answered. OPEC cannot afford the luxury of having an unclear vision of the future or erring in its strategic choices, as clearly happened to the oil companies in 1959.

Reflections and Questions

Making the decision in 1959 was difficult, because the producers feared loss of their markets by confronting the powerful oil companies. However, there were strong reasons for an easy and quick consensus. They had to join forces to defeat a group of "foes" in the form of the international oil companies. There was an absolute clarity of objectives. On the one hand, it was evident that the producing countries were receiving a very small portion of the benefits of oil exploitation, so their demands were well-founded. On the other hand, by controlling 80% of total oil exports, the OPEC countries wielded immense power, and held the upper hand in any possible dispute. Once the decision was made, the path became crystal clear. However, the situation nowadays is considerably different. OPEC's market share is around 34%, and the quota system has proved to be more complicated than anyone would have imagined. There are important reflections and questions to be answered. What are the strategic issues confronting OPEC in today's world? Who and where are the "foes"? What is the right price target, and what are the considerations that should guide the definition of this target? Is the current price band appropriate? Is there a reasonable way for OPEC countries to balance their fiscal needs with the price and volume objectives? What about the dilemma of market share versus price? What are the expectations of individual members concerning their quotas? Under what price conditions would substitution gain importance and threaten oil as a source of energy? These are the issues that need to be addressed.

The Substitution Question

In 1865, Williams Stanley Jevons, one of the most distinguished economists of the Victorian period, published what was to become his most famous study, *The Coal Question.* In his work, Jevons established a clear connection between the rise of the British Empire and the massive use of coal as source of energy to feed the steam engines. At that time, projections showed that England had coal for 2000 years. Yet, a few years later, the forecasts were not so optimistic, and those 2000 years were reduced to just 200 years. Jevons even predicted that the exhaustion of coal would mark the demise of the British Empire, since in his own words "none of the other sources of energy can replace its supremacy."

One and a half centuries after Jevons, Great Britain still has enough coal reserves to cover its needs, and this coal can be exploited at the lowest cost in the world. Yet, it is more important to realize that coal supplies less than 10 percent of the total energy needs of the country, since most of it has been replaced by oil and natural gas. However, history shows that substitution never translates into total disappearance. For example, wood was replaced by coal and oil, while copper was replaced by optic fiber, but both wood and copper are still being used on a small scale in their particular niches. Other commodities, including natural fibers such as cotton, silk and wool still retain their own niches, despite having been largely substituted by artificial fibers.

Since its first discovery by drilling in 1859, giving birth to a new industry, the production of oil has increased all around the world, having reached 80 million barrels per day (mbpd) nowadays. According to any credible forecast, oil will continue to be the fuel of choice for many years to come. However, an important current discussion deals with the question of how long the oil reserves can last. There are several schools of thought on the issue, but they can be grouped into two major streams. The pessimistic ones believe that oil production will peak soon. Their view is clearly expressed by the work of Colin J. Campbell and Jean Laherrère, called *Supply of Oil 1930-2050*, based on a cold geological-arithmetic calculation. They conclude that oil production will peak in about a decade, and restrictions will soon begin to be felt all around the world, with the exception of the Middle East. By 2050 oil production would have dropped to some 18 mbpd, the level of 1950. In other words, this scenario revives the traditional Malthusian theory. Based on these predictions, many others are estimating catastrophic consequences for civilization, irrational use and exhaustion of other natural resources, irreparable damage to the environment and an imminent worldwide social crisis.

The optimistic view, in terms of the benefits for the world, but not necessarily of the benefits for the oil industry, comes from those who believe in the immense impact that competition in the markets, technological development and human ingenuity have on production. Along those lines we find economists such as Robert Hirsch, Thomas R. DeGregori, Julian Simon and Morris Adelman, who believe that the fact that the earth's crust contains finite amounts of a mineral is irrelevant, because the exploitation of that mineral will cease long before it is exhausted. As exploitation costs increase, the substitutes will become more competitive, that industry will decline, and an undetermined amount of that mineral will be left in the earth, although exploitation may continue at highly reduced levels. Cost here does not refer to exploitation alone, but refers to relative costs among substitutes, including taxes, regulations and environmental restrictions. In other words, the necessary production will always be available. It is what Professor Vaclav Smil, of the University of Manitoba, has labeled "adaptive cornucopia."

Concerning oil, the costs of transforming existing resources into proven barrels does not give any indication of imminent exhaustion. With at least 1 trillion barrels of proven reserves and a supply curve showing the marginal barrel at around $15, it is fair to assume that oil will continue to be a dominant force for many years to come. Besides, many years will pass before substitutes become competitive in the transport sector, to displace oil in large amounts, despite the growing incentives. This means that we can anticipate the addition of new reserves for many years into the future. However, there is a caveat to this affirmation. If sustained for a considerable time, high prices of oil will have an effect equivalent to high costs and will undoubtedly accelerate substitution. There is abundant statistical information leading to the conclusion that when prices remain long enough at or above $30 per barrel of West Texas Intermediate (WTI), oil market share starts falling from the level of ~ 40% that it has sustained for a long time. So a note of caution must be issued for countries with large oil reserves. If oil is to remain the fuel of choice for many decades, the target price will have to be competitive. The level of ~ $40 per barrel observed in recent months, is a bad recipe for the long term. A final reflection on this point is that since supplying the world with a growing demand for oil is not an objective in itself for the actors that supply it – international oil companies (IOCs) and national oil companies (NOCs) – there is a risk that supply may falter during the medium term, with severe market consequences that would affect all, but with the greatest loser being oil.

Nationalizations and their Impact

With the exception of Mexico in 1938 and Iran in 1954, nationalizations took place in oil-producing countries around the 1970s. Those countries could have continued without changing the scheme that had prevailed for decades—the oil companies producing in their territories battling with the slim margins permitted, while the governments collected royalties and taxes to be used in their socio-economic plans. However, OPEC members were under great pressure from public opinion and were bound by their own history. There had been too many years of exploitation and poverty. To expel the companies was a moment of bliss, yet they soon began to pay the price for it, and according to many the cost has been too high. My opinion, based on the Venezuelan experience, is that nationalization was a necessary step. In a country where oil is so predominant, it would have been impossible to have a stable long term system, based on a purely fiscal link to the country's oil industry, with the government limited to collecting the rent, while the country remained ignorant in oil matters. The reference to ignorance is not intended to be pejorative, but the reality is that the oil countries did not have deep knowledge of the complexities of the international oil industry, and those nationals who could be considered experts were often employees of the oil companies. In the Venezuelan case nationalization took place in 1976. Thus, in 1994 when Venezuela decided to open up the oil industry ("Apertura Petrolera") on a large scale, this transcendental step was supported by a mature, efficient, modern, well-managed and technologically up-to-date national oil industry, with qualities directly resulting from the nationalization. The national oil company, PDVSA was financially strong, with an ample international presence and was managed by professional people, with proper interlocutors in all corners of the world. Under those conditions, it was possible to ensure a well-balanced and appropriate opening process.

To return to our story, the nationalizations detached the upstream (exploration and production) from the downstream chains (refining and marketing), creating separate open markets for oil and for refined products. OPEC producers had to create new marketing platforms, and mostly began selling oil to refiners that were operating with very small margins, susceptible of being squeezed or even inverted by any small price variations. This originated in swaps, futures and other stabilizing mechanisms, used by buyers to protect themselves from volatility. The buyers would seek the best offer and then hedge their volumes in the futures markets. Thus, ironically the futures markets were a consequence of the nationalizations in the oil producing countries, principally OPEC members.

The Impact of the Years 1982–1983

The years 1982 and 1983 were difficult and crucial for the future of the global oil industry. The breach between the international oil companies and the oil exporting nations, particularly the OPEC members came to the fore once again. Prices had escalated sharply in 1970, 1973 and 1979 and the world entered the 1980s with very high oil prices. These prices triggered campaigns of energy savings and increased energy efficiency in the industrialized countries, and permitted new and more expensive production developments—the North Sea with 5 million bpd being perhaps the most notorious case. As a result, prices began to show signs of weakening, and in an emergency meeting in March 1982, OPEC decided for the first time to cut production (700,000 bpd), in an attempt to sustain prices. It was not the first time that the issue had been discussed, but previous attempts had been unsuccessful. OPEC's decision provoked many discussions. For example, the Group of 30, a high level body of international economic and monetary experts and government officials, warned about the effects of very low prices on oil demand, and favored a tacit collaboration between exporting and importing nations. The Brandt Commission on International Development Issues called for a dialogue between major oil producers and oil consuming countries, to consider arrangements beneficial to all parties, including safeguarding supplies to the poorest countries. Many experts rendered opinions and proposals. Sheikh Ahmed Zaki Yamani played an important role, by heading an OPEC ministerial commission. He showed that he was better informed than most, with his idea that prices should be lower for OPEC to remain in control, but at the same time warning consuming countries that if prices fell too low, there would be bankruptcies all over the world and the banking system would come under serious pressures. Most oil and gas loans had been made on the premise of a continuous increase of energy prices.

The importance of this retrospective is to underscore the contentious character of the relationships between oil companies and producing nations, to distinguish the first attempts of cooperation between producers and consumers, and finally to remember the birth of the OPEC quota system that would become formal in 1983.

New Evolution of the Pricing System

Despite all the efforts made during the early 1980s, oil prices eventually collapsed in January 1986. However, the pricing system had begun to change for good. The posted (notified) prices began to disappear. For a long time, the large volume had been traded at notified prices, while only the marginal barrel was traded in the spot market. Progressively all operations began reflecting the spot prices, and later on came the dominance of the futures markets, since these are a transparent and widely known proxy of the spot prices. Before the development of the futures markets, the small oil companies and the independent refiners were in the hands of the large suppliers and their pricing conditions. Today those small companies are in the optimum position to bargain with suppliers, because prices are public and transparent.

The Futures Markets

Futures markets involve a contractual obligation by both parties to either make or take delivery of the underlying commodity at a specified future date and at a specified price. Cases of physical delivery are always lower than the contracts traded. In fact, less than 2% of all futures contracts traded actually result in physical delivery. In 98% of the cases, the contractual obligation is settled with an "offset." This means that the obligation is satisfied by selling a similar contract in which each party assumes the opposite position to its original position. That allows for the two positions to offset each other, and thus the obligation to be satisfied. Naturally, the trader assumes financial responsibility for the difference in price between the two positions. The existence of a ready and liquid market to buy or sell futures contracts, allows commodity futures markets to act principally as financial markets. The exchanges are public in the sense that anyone who makes the necessary arrangements with member brokerage firms can trade. The offset option means that nobody needs to have physical volume in order to trade futures.

There are two parties to every contract – one party agrees to buy and take delivery in the future and the other party agrees to sell and make delivery in the future. The buying party has what is referred to as a long position. This party is called the buyer and is said to have bought a contract. The selling party has what is referred to as a short position. The short is called the seller and is said to have sold a contract. A long position makes profits if prices increase and incurs losses when prices decrease. The short position is the opposite—profits occur when prices decrease and losses occur when prices increase.

Energy futures and options contracts and markets provide benefits to market participants and the public at large. One of the most important benefits is a highly visible and efficient price discovery mechanism. Futures contracts trade under conditions that closely approximate perfect competition—there are a large number of buyers and sellers conducting transactions of a homogeneous product. Exchange rules ensure fair, open and honest trading. In addition, there is an efficient and essentially costless information system, and except for the financial responsibility there are no barriers to entry into the market. Under such conditions the price that results can be considered the "proper" price, because it represents the consensus resulting from the decisions of many thousands of independent market participants.

Prior to the advent of the futures markets there was no visibly, widely quoted benchmark price. Often those prices quoted were subject to modifications and discounts which rendered them useless in making decisions. OPEC posted prices were often subject to political as well as supply–demand considerations. Different major oil companies would often sell product to their customers at substantially different prices. There were also different quotations in different geographical locations.

Futures markets prices are not subject to the problems of posted and spot market prices. Futures are determined in a central marketplace and are representative of standardized quantities and qualities of petroleum product trading at a particular location (the delivery point of the contract). Finally, futures markets provide signals that help the allocation of product over time and tend to smooth out seasonal variations of supply and demand in the cash market.

The Price Collapse of 1998: Origin and Consequences

Between 1994 and 1997 the demand for oil increased steadily at an average rate of 1.5% per annum. In 1997 the increase was of 2 million bpd. This increase provided the scope for additional volumes from every producer worldwide, including OPEC countries, and it is fair to say that within the organization, discussions about quotas and ceilings had turned sour. At the center of the dispute were Saudi Arabia and Venezuela. One of the underlying factors of the controversy was the fact that Saudi Arabia had absorbed two out of the three million bpd of Iraqi embargoed volume, while the remaining one million had been distributed among the other 10 members. Another factor was that Venezuela was reluctant to accept a quota of 2.325 million bpd that it considered disproportionately low and inconsistent with its fiscal needs and investment plans, particularly in heavy/extra-heavy oil.

During those years, despite many indications of fragility of the economies of Southeast Asia, the world continued to believe in "the Asian miracle." Not even the collapse of the Thai economy in August 1997, which would eventually trigger a domino effect in the whole region, received sufficient attention. Thus the world continued to be optimistic about economic growth, unaware of the abyss that lay just around the corner. As a result, the consensus projections of oil demand growth for 1998 were around 2.1 million bpd. Within that landscape, OPEC's ordinary conference took place in Jakarta in late November 1997. The decision of the ministers to reveal the actual production levels (and not to increase production, as has been suggested frequently but wrongly) may have been a psychological factor in the price collapse of January 1998, but it certainly was not the cause, since it was not structural in nature. The real reason for the market shock was the collapse of the Asian economies, and the worldwide impact that followed, translating eventually into an increase in oil demand in 1998 of only 200,000 bpd, a shortfall of almost 2 million BD.

The necessary adjustments were very difficult to address and implement. In early 1998, Saudi Arabia, using Mexico as intermediary, approached Venezuela to try to work out a joint solution. A meeting of the three parties held in Riyadh in March 1998 was the first step to coordinate new production quotas in an effort to balance the market, while prices had dipped to $10 per barrel. Two more meetings of the group of three, one in Amsterdam in May and one in Madrid in December, combined with two OPEC ordinary conferences, plus many extraordinary meetings and consultations, gradually began to restore stability to the market. It is important to mention, that the process of year 1998 was of trial and error, because the real magnitude of the crisis was not evident until the second half of the year. By January 1999, the oil price had reached some $18 per barrel and it was on its way up, since the global economy was beginning to rebound and the supply of oil was in balance with demand. The crisis was deep enough to soften the dissidence that had prevailed for several years. In February 1999 a new government took office in Venezuela, which became a price hawk and, operating in better harmony with Saudi Arabia, contributed significantly to the price increase that would come about during 1999–2000.

2002–2004: The Misleading Years?

The very high prices of oil observed in the market have made many people wonder whether the world might be entering an era of oil scarcity. However, we have already established the premise that the world is not running out of oil. Why then is the price of oil at $40 per barrel? A quick look at the market will provide some answers. In a normal market when demand grows, inventories fall, prices go up and that will trigger additional supply. However, the markets are currently locked in a self-reinforcing loop of low inventories and high prices. How has this situation come about? For an explanation, let us turn briefly to the past, to January 2002. The consensus forecasts for the price of oil in 2002 were around $21–22 per barrel of WTI. It is important to recall, that OPEC had agreed on a production ceiling of 21.7 million bpd beginning on Jan 1, 2002 (added to some 2 million bpd from Iraq, it amounted to 34% market share, OPEC's lowest since 1980). However, the actual average price at the end of the year was slightly above $27 per barrel of WTI. Several factors of uncertainty affected the psychology of the market—threats of an oil strike and political instability in Venezuela, the fear of a Middle East embargo in rejection of the US management of the Israel-Palestine conflict, and the threat of a war in Iraq. Although the first two factors were short-lived, the third not only lingered, but also intensified during the year. Because of the resulting "war premium" (initially estimated at some $5 per barrel), an extended backwardation set in, which provoked counter-seasonal stock movements in the second and third quarters. During the second quarter, drainage of some 400,000 bpd occurred, instead of the normal stockpiling of 1.3 million bpd, and in the third quarter, drainage of some 300,000 bpd took place, instead of the normal stockpiling of 400,000 bpd. As a result, when demand began increasing in preparation for the winter, oil inventories were low (in the United States they had crossed the symbolic threshold of 300 million barrels, and were fluctuating between 270 and 280 million barrels). This brought increased pressure on prices, which remained high despite OPEC exceeding its production ceiling by almost 4 million bpd. Two upward adjustments of that ceiling were not enough to diffuse the tightness of the market, particularly because of the stoppage in Venezuela in December. Then the war started, but being short, it raised hopes of Iraq's production returning to normal quickly. However, extended sabotage crippled production capacity, and restoring it has taken much longer than expected, keeping the market tight. Also, Venezuelan production never came back to the pre-strike levels (only 2.5 mbpd instead of 3 mbpd), and there have been threats of stoppages in Nigeria, which have also contributed to upward price pressures.

The Backwardation Vortex

There are times when the market price structure is "backwardated" to the point that it prevents inventories from building up and the low stocks create even more backwardation, such that the cycle feeds on itself. This situation has historically been described as a "backwardation vortex" by Energy Security Analysis, Inc., and has typified the market dynamics over the past 26 months. The "vortex breaker" would be the development of a prompt crude overhang or a big decline in demand, or both.

Supply has been increasing significantly, and the stocks have recovered progressively (already above 300 million barrels of oil in the United

States). However, demand has continued to strengthen despite the high prices (above 2 mbpd in 2004 and 1.8 mbpd in 2005). However, when it looked like the fundamentals were catching up, the terrorist attacks in Saudi Arabia (although production was unaffected), new production interruptions in Iraq and the Yukos affair, increased the psychological premium in the market. By mid-August, prices had peaked at around $48 per barrel, with some analysts estimating that premium as high as $18 per barrel. For a long time now, except for the first three months, the forward curve remains in severe backwardation.

The Organization of Petroleum Exporting Countries

Quoting the Petroleum Intelligence Weekly (PIW), the vortex has allowed OPEC to play "smoke and mirrors," and enjoy the double benefit of price and volume. Certainly, there is nothing wrong with that. On the contrary, it has been good for OPEC producers. However, it is only a matter of time for OPEC to be forced to revisit the dilemma of high price vs. market share. It must be kept in mind that as a result of OPEC seeking a higher price, for several years now the average annual increase in demand of some 1.0–1.4 million bpd has been taken entirely by non-OPEC producers, especially Russia. This has also translated into pressure for quota increases coming from Nigeria, Algeria and Venezuela. The best way of expressing the above-mentioned dilemma is to look at the numbers calculated by the Energy Information Administration (EIA) of the US Department of Energy (DOE) for diverse long-term price scenarios (see Table 2.1).

As shown by the numbers, the difference in volume between the cases of high and low price, represents for OPEC a differential volume in 2025 of 21.7 million bpd. In my view, in the long run, without abandoning its role as swing producer, OPEC will have to settle for a more moderate level of prices. This however is easier said than done, because of the fiscal needs of the different members.

Table 2.1
Long-Term Price/Production Scenarios

Price in 2001 ($ per barrel)	*33*	*25*	*18*
Production (mbpd)			
OPEC in 2015	34.3	41.4	48
Non-OPEC in 2015	59.6	57	54.1
OPEC in 2025	45.2	55.6	66.9
Non-OPEC in 2025	67.8	62.8	58.1

Source: Energy Information Administration (EIA), US Department of Energy (DOE).

OPEC's Objective: Highest Price or Long Term Viability?

It is natural that the seller of a product will welcome the highest price possible. However, common sense dictates that the owners of large oil reserves should be interested in ensuring an optimum market presence for their oil for many years to come. In other words, suppliers cannot consider only the supply side, but must also be concerned with economic activity and oil demand, elasticity of demand, potential substitution of oil, and technological and environmental trends. This broad view will allow for optimum planning, aimed at ensuring a larger market share and long-term sustainability. In a publication called *Fuel for Thought* the World Bank presents a summary of long-term scenarios of how the slate of energy sources will look in future global consumption. The above-mentioned summary is drawn from scenario planning done by the World Energy Council (WEC), Shell, Inter-Governmental Panel on Climate Change (IPCC) and other sources. While acknowledging that "oil is more abundant than anyone had thought, and its convenience as a transport fuel is likely to guarantee it an important place in future consumption," the World Bank also concludes that "renewables could provide up to 50% of all energy by the mid-21st century, given appropriate enabling policies and new technology development."

If the OPEC mantra is presumably that producers should join forces to make the best from the market in the medium–long term, the pricing policy is of crucial importance. The common discussion about "fair" prices is irrelevant. In economics there is no such a thing as fair price. The price target should be selected with the purpose of maintaining oil as king in the energy market for a long time, and keeping non-OPEC producers at bay. Nevertheless, OPEC's decisions on prices have been increasingly influenced by the financial needs of member states. The higher their expenditure, the more urgent is their need for higher oil revenues. As a result OPEC's primary inclination is to try to increase prices. By now the market believes that the organization will seek higher and higher prices, and that the $22–28 per barrel band is past history, although that may not be the case. In fact, for several months now prices have been above the upper limit of the band, and instead of making any reference to the importance of staying within the band, there are calls from OPEC members for an increase of the band prices.

As a reference, on the basis of a calculation by the Centre for Global Energy Studies (CGES) for the year 2004 and considering a production level of 8.2 million bpd, Saudi Arabia needs a price of $23.2 per barrel to cover general expenditure, $26.7 per barrel to add debt service, and $32 per barrel to add retirement of some $4.5 billion of debt. Another example, perhaps more dramatic is that in 2004 Venezuela would require nearly $50 per barrel to balance the government's expenditures. However, can this possibly be the right way to go? A price as high as $40–45 per barrel, observed in the past 2 years, if sustained in the long run, will effectively be equivalent to the high cost referred to when addressing the issue of alternative sources. It will prompt a faster substitution of oil. A Hubbert-type analysis of global oil production since the beginning of the 20th century, compared to actual output, shows how dramatically the latter departed in 1979 from the theoretical curve. This was precisely the result of attenuation of consumption by energy efficiency and substitution (demand destruction), which were direct consequences of the very high oil price.

The Corporate Environment

Until only a few years ago, the market would value oil companies on the basis of large volumes and deep pockets. However, in recent times a major shift has taken place, with the market looking closely at performance, a balanced portfolio and return on capital employed. Emphasis is shifting from revenues to profits, from share of market to share of value, from assets to people, from operations to customers and in general from the inside to the outside. These changes reflect the search for a new business design since, during the last 20 years, the performance of the oil sector and the energy sector in general have been lackluster when compared with other industries.

Oil companies face the challenge of competing with other industries for the capital they need for consolidation and growth. This trend, coupled with the trend in oil-producing countries of expansion and opening to private participation, allow predictions of a future oil world more integrated and efficient, as well as much more competitive.

In summary, companies are making a swift transition from being oil and gas companies operating in the oil and gas domain, to global businesses with a strong commitment to sustainable growth and development.

British author John Elkington describes this very well in his book *Cannibals with Forks*, which defines the concept of a "triple bottom line" as an imperative for success in the modern corporate world. The concept states that, in addition to economic return, the bottom line has to include environmental quality and social justice.

That certainly poses a big challenge in itself. Nowadays, matters have become more difficult as a result of the corporate scandals of the last few years. The case of Enron and several others have dramatically complicated the corporate landscape. Companies today have to adapt to stiffer legislation, such as the Sarbanes–Oxley Act and the 1998 Combined Code on Corporate Compliance in the United States, the Code Tabaksblatt in

The Netherlands and others. They also have to deal with better informed, more inquisitive and less tolerant shareholders, who have learned a lot about the markets, and are willing to take their money somewhere else. They are facing difficult decisions to maintain the proper balance between total shareholder return (TSR) which includes dividends plus share price, reserve base and reserve replacement, future growth and financial strength. A few relevant points may be noted below:

Reserves: Practically every major oil company has enough reserves to sustain oil production for some ten years and production of gas for some 15 years. Here, an exception should be made with BP, as a result of its recent Russian acquisition (some 35 billion barrels of oil equivalent). Better technologies in reservoir engineering and production geology alone will not be enough to grant a significant production life extension. Only mergers, joint ventures and acquisition can deliver such an extension.

Buy-backs: The recent buy-backs have not generated a significant positive impact, and in the end have not been received well by the investment community. In general, it is felt that this is not the best use of cash.

Credit-rating: The credit rating is very important for most of the oil majors. However, the big question is when, if ever, and under what conditions it may be worthwhile to sacrifice the AAA rating in order to ensure future growth. A debt to equity ratio of 30% plus is roughly the threshold for maintaining the AAA rating.

Return on average capital employed: Big Oil's capital spending is rising as it struggles to replace reserves and develop long-term projects for future decades. This is putting the companies return on average capital employed (ROACE) under threat, much to the dismay of investors, who have been encouraged to use this measure to assess company performance.

Capital discipline: Majors are now trying to offer reassurances to investors on overall capital discipline, while easing self-imposed restrictions on investment. One symptom of this is the use of less conservative oil price assumptions, according to the Petroleum Intelligence Weekly (PIW).[1] Another symptom is a less religious adherence to ROACE targets. BP, for example, has sunk $7.7 billion into Russia knowing that it will not achieve its target of a 13% ROACE at $16 per barrel of Brent crude. Chief Executive Officer John Browne has since admitted that he considers ROACE "really a quite dreadful measure," poorly defined and virtually useless for internal planning purposes. Royal Dutch–Shell, too has said it will not adjust its ROACE targets to reflect external factors such as accounting changes and foreign exchange effects, and redirected investors instead towards its underlying cash management. "Ultimately, normalized ROACE will fall where it will, depending on factors outside our control. We manage according to the underlying cash," a Shell investor relations official said recently.

Total returns: Having been weaned off targets first for volume growth, and now returns, investors will want something else to consider. BP's John Browne has suggested that his executive team should look primarily at total returns – share price plus dividends – but some investors think that this also offers a poor guide to management performance. In the words of a UK fund manager, "We need some way of assessing whether they are delivering against what they've promised, but I've never been a fan of total shareholder returns. The company actually has limited impact on its share price." Oil company management cannot take the credit for oil prices close to $40 per barrel, for example, but this is arguably the main factor that has sustained share prices and overall returns to shareholders during 2003.

Softening targets: Many analysts, however, think the oil majors are taking a step in the right direction by softening ROACE targets. Moreover, by committing to ultimately unrealistic goals, Big Oil has set too high a bar for itself when making investment decisions, consequently missing out on attractive opportunities.

National Oil Companies

The basic principles described above are governing the world of international oil companies (IOCs). Nevertheless, the description of the oil and gas landscape would be incomplete if National Oil Companies (NOCs) are not put into perspective. NOCs have been around for many decades, but their importance has grown currently to very large proportions. Saudi-Aramco, Statoil, NIOC, KPC, PDVSA, Pemex and Petrobras, among others, underscore that importance.

While the IOCs have constantly been in a process of reform and change in their organization, financing strategy, technological base and investment screening criteria, NOCs have not kept up with the pace of progress and are lagging far behind. A revision of their numbers shows, that if NOCs such as Aramco and KPC had been publicly traded, their stock prices would have crashed. To a great extent, this is the result of the differing goals between the NOCs and their respective governments (or shareholders).

The most important goals of NOCs are to:

- Develop their resources
- Apply the best technologies
- Expand production
- Maintain business
- Grow organically and profitably

On the other hand, the goals of the governments concerning their NOCs are to:

- Increase domestic fiscal revenues
- Increase the country's export earnings
- Achieve social goals
- Achieve political goals

These differences must be reconciled if NOCs are to be competitive in an environment where efficiency, better returns and clean energy are key factors for attracting capital investments in new developments.

Although there is no "black-and-white" recipe because there are many differences among NOCs, the higher successes have come to those NOCs belonging to countries that have managed to establish a clear separation between policy, regulation and operations. Norway and Brazil, for example, place policy in a Ministry, regulation in a Special Agency and operations in the National Company.

In some cases, such as Spain and Argentina, the process of growing participation of private capital has led to total privatization (Repsol and YPF).

In countries where the oil and gas activity plays a commanding role in the economy, it is difficult, perhaps inconvenient and even impossible, to conceive total privatization. But the above-mentioned separation of policy–regulation–operations will allow for much better NOC functioning and for increased participation of private enterprise in operations, without compromising critical government issues (or dogmas). The end result will be steady growth in the oil and gas sector.

Although for many years the potential for complementation between IOCs and NOCs has been recognized, there has been very modest progress in taking advantage of the hundreds of win-win opportunities available. However, it is clear that not doing it means turning your back on large value additions, and it is becoming clearer that it may be the only way forward, as both IOCs and NOCs look into the future.

Summary and Conclusions

The creation of OPEC was a transcendental decision, not only for member countries but also for the global oil industry. It showed great courage and most importantly a clear futuristic vision. Almost 4½ decades later, the organization continues to play a major role in setting oil prices, and as a result, the prices of other energy sources. During that time, though, many things have changed. The price mechanism of posted prices, once in the hands of the major oil companies and later on in the hands of OPEC, belongs to the past. Although OPEC's influence on the price levels is a determinant, it is indirect, and the futures markets constitute the pricing mechanism for every barrel of oil traded today.

For a long time the interests of oil importing and exporting countries have been perceived as very different, and in many cases antagonistic. The exporters want high and constantly rising prices, while the importers want steady supply at moderate prices. These diverging perspectives have lost their relevance. As mentioned, in contrast to what happened up until the 1980s, neither exporters nor importers set the price of oil, an issue that has perhaps been the most contentious over the years. Therefore there appears to be new fertile ground for cooperation.

OPEC's cohesion has often been shaky over the years, mostly as a result of quota struggles, which are generally related to specific individual revenue targets. But the thorniest issue relates to price targets and market share. The appetite for ever higher prices may eventually become the vulnerable spot or the "Achilles heel" of OPEC, because maintaining the price within the OPEC band has implied relinquishing market share systematically to non-OPEC producers. It is important to clarify that the market behavior during the past 28 months is an anomaly that has remained for so long that it seems normal. It conceals the existence of this problem, and may be very misleading. This prolonged period of very high prices may have blurred our vision and will not help the case.

A second problem derived from high prices is of greater importance, though it relates to the longer term. It concerns oil maintaining its predominance within the energy slate, which implies targeting competitive prices. This is a problem that in the longer run will become compounded with substitution trends, which in the case of renewables, have their own dynamic inertia beyond inter-fuel price competition. It is extremely important to carry out modeling of future trends, in order to optimize timely exploitation of oil resources, and delay substitution as far as possible into the future. Proven reserves of over 1 trillion barrels of oil and 5,600 trillion cubic feet of gas indicate that no scarcity is expected, but if some credit is given to the Cornucopians, we should keep in mind the cases of copper, silk and cotton, if not coal.

Concerning market share, the proper pricing policy could give OPEC an advantage of 20 million bpd over non-OPEC producers by 2025.

The national oil companies will not be able to address all the future opportunities alone. The international oil companies, especially the oil majors, are centers of technical, managerial and financial excellence. They can provide everything needed in terms of technology, financing and optimum access to markets, and are the natural complement for the needs of national oil companies. It would be foolish for oil producers to reject the opening of their upstream industry. A strong upstream industry is also the best bargaining tool to acquire reliable access to the more attractive markets, since the most pressing issue for international oil companies is access to new reserves. Dozens of different deals can be worked out, always ensuring the lion's share for the producing countries. In this context, it is important to mitigate the risks to the investors, especially the tax and fiscal risks but more importantly to ensure that governments in the producing countries remain guardians of the law and respect contracts and technological property.

It is highly recommendable that the OPEC countries start working on the implementation of governance schemes that guarantee the separation of policy, which is the responsibility of the government; regulation, entrusted to a special agency, as it has been the case in Brazil and Norway; and operations, being managed by the national oil company. In my opinion, this is the only way to ensure the necessary efficiency and least interference.

In 44 years since the foundation of OPEC, with a few minor exceptions of joint enterprise between Arab countries, there has not been a single oil and gas joint venture between or among members of the organization. If such a venture becomes a reality – and I am certain that many options exist – it would convey a strong signal of unity to the whole world. OPEC is not perceived as a group of partners, much less as a team. It is viewed simply as a heterogeneous group with little in common, which meets on occasions to try to increase the price of oil.

The fourteenth century Arab philosopher, historian and sociologist Ibn Khaldun once said, "The commercial activity in charge of a ruler is damaging to his subjects and ruinous to the fiscal accounts." Although this statement may sound too strong, it is well accepted that a larger constituent base will always offer better accountability. State companies, even if they are not corrupt or extravagant, encounter difficulties in undertaking efficient investment programs. The money they manage, as allowed to them by their respective governments is usually challenged by other national needs and priorities, often leading to paltry investments. It seems that OPEC nations have for a long time recognized the convenience of an organized return of oil multinationals to their territories to participate in exploration and production (E&P) ventures, bringing with them capital and technology, and opening up new market avenues. The multinationals are always willing to assume risks, provided there are prospects of gaining attractive rewards. That is precisely the path that Venezuela has tried to follow. Unfortunately, this effort has stagnated, and in some aspects has been reversed under the current government. It must be admitted that for producing nations it has been very difficult to transform that plan into specific actions. It is a real pity that oil producing nations today are still exhibiting the spirit of the mid 1980s. State oil companies and tax systems remain hurdles in the global effort to open up and develop new prospects, in an oil-thirsty world where environmental trends are accelerating, and the window of opportunity for fossil fuels may be narrowing. Despite its current political problems, Russia has illustrated what can be achieved when restrictions are removed and market forces are unleashed.

On a more personal note, I would like to make a few observations. As a member of the Venezuelan delegation, I had the privilege of attending OPEC meetings regularly between March 1994 and January 1999. However, after I stepped down as President of PDVSA, a position reached after a 30 year-career in the oil industry, efforts were made by certain Venezuelan officials to distort the purpose and outcome of my work. This has unfortunately led to the misconception that I am personally opposed to

OPEC and to countries in the Middle East region. This is absolutely false and I would like to set the record straight. I have great appreciation for the professionalism of all the OPEC delegates and count many friends within the organization with whom I share great memories. I was an active participant in the difficult initiatives of 1998, intended to restore balance to the oil market after the collapse of the economies of South East Asia with its global ripple effects, which was the real cause for the plunging of oil prices in January of that year. My arguments and discussions within OPEC were always inspired by the best interests of my country, and were directed towards enhancing the collective benefit of the members.

Dissension and argument should never be interpreted as disloyalty. To quote contemporary American businessman William Wrigley Jr., "When two men in a business always agree, one of them is unnecessary." This clarification is necessary, because without intending to be disrespectful, I want to conclude this analysis by pointing out that in my opinion OPEC meetings are often tense and shortsighted. It is not enough to look at the supply-demand-stocks numbers, split up to calculate individual needs separately, and then come back to put together a mosaic that can keep the situation under control for one or two quarters. It is my firm belief that the future long-term success of OPEC requires frank deliberation on all the issues discussed above. This implies bridging the trust gap. Although there are good friendships among the OPEC member states, there seems to be a strong collective sense of mistrust and sometimes formal and excessive diplomacy at the core of the organization. Flushing out that collective sense of mistrust could perhaps be the first step in building a new and modern OPEC, which can successfully meet the challenges of the times.

Section 2

ENERGY CASE STUDIES

3

Iraq's Oil Sector: Scenarios for the Future

Muhammad-Ali Zainy

In the aftermath of the tragic events of September 11, 2001, a new reality has dawned on the world's oil consuming countries—given the identity of the perpetrators, Saudi oil supply may not be as solidly secure as it once appeared to be. However, Saudi Arabia is home to a quarter of the world's proven oil reserves and holds most of OPEC's spare production capacity. As such, it is effectively the world's swing oil producer in case of an abrupt increase in world oil demand, and also the world's shock absorber in case of disruption in oil supplies from another producer.

The question that arises is, how can the world's oil supply security be assured if Saudi Arabia's position as the leader in world oil reserves and production capacity and the lynchpin behind the security of world oil supplies may itself be insecure? This was one of a host of reasons, whether real or imagined, that prompted the United States principally, and its allies – notably Britain – to decide to topple the regime of Saddam Hussein. Among the other reasons were: removing Iraq's weapons of mass destruction (WMDs); crushing one of the world's worst dictatorships; ridding the Middle East of an incessant source of instability; and perhaps, democratizing the whole region.

America's calculation was correct insofar as Iraq's oil was concerned, and in light of the world's new oil supply realities, to the extent that the

Iraqi oil frontier could be opened and its massive reserves could be tapped. The whole world turned its attention to Iraq and the possibility of benefiting, in one way or another, from its enormous oil riches. Oil companies, large and small, with high expectations, prepared to plunge into Iraq at the first opportunity to participate in the development and exploitation of the country's ample and low-cost oil reserves. Oil analysts started forecasting optimistic scenarios of rising Iraqi oil production, with a widely accepted scenario of Iraq regaining its pre-Gulf War sustainable production capacity of 3.5 million barrels per day (mbpd) within two years of toppling the Iraqi regime, and possibly reaching a capacity of 6 mbpd, or even 8 mbpd, within 5–6 years of toppling the regime.[1]

By 2005, more than two years after the collapse of Saddam's regime, none of the above-mentioned expectations seemed likely to materialize within the time frame envisaged. Indeed, Iraq has failed to deliver. The following section analyzes developments, presents likely scenarios that Iraq's upstream oil industry may assume, and pinpoints the necessary conditions for Iraq to regain control of its oil riches and expand production to levels commensurate with its oil reserves. Only if these conditions are fulfilled can Iraq provide the rest of the world with the secure and ample oil supplies much needed by consumers now and for the foreseeable future.

Iraq's Oil: An Overview

The first oil discovery in Iraq, the super-giant Kirkuk field, was made in 1927, just two years after granting an oil concession to the Iraqi Petroleum Company (IPC). Twelve years later, in 1939, the Ain Zaleh field was discovered in the Mosul area by the Mosul Petroleum Company (MPC), and ten years later, in 1949, the giant Zubair field was discovered in the Basrah area by the Basrah Petroleum Company (BPC). According to the three concessions granted to these companies, the IPC controlled the entire territory north of Baghdad and east of the River Tigris, the MPC controlled the area north of Baghdad and west of the Tigris, while the BPC controlled the southern part of Iraq. As such, virtually the whole of Iraq came under the concessionary control of these three companies (see Figure 3.1).

Figure 3.1
Petroleum Map of Iraq

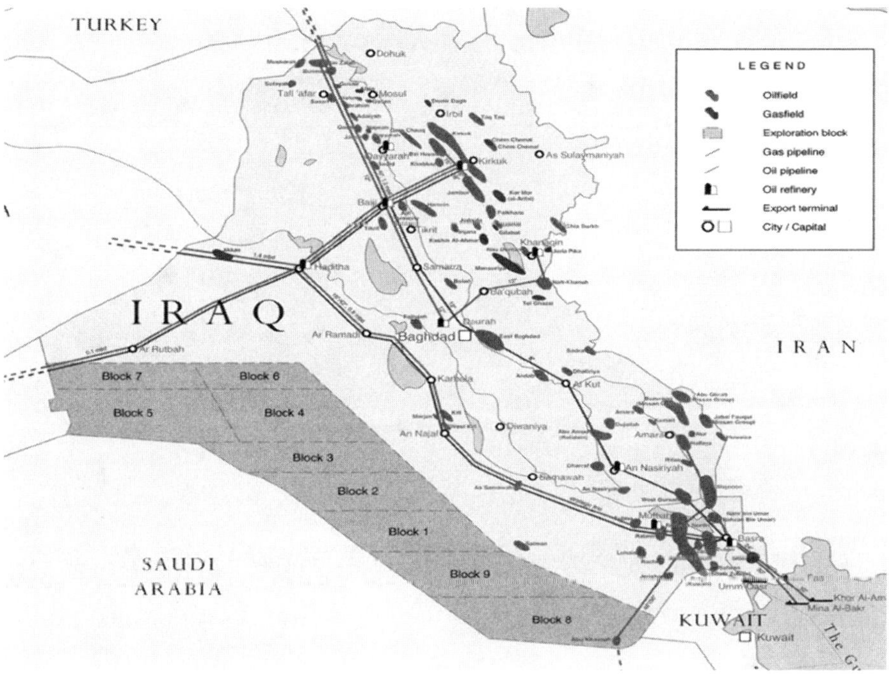

Source: Centre for Global Energy Studies (CGES), London, UK.

A Period of Stagnation

The three companies IPC, MPC and BPC were principally controlled by the same shareholders: the Anglo-Iranian Company (later BP), Royal Dutch Shell, Compagnie Française des Petroles (CFP, later TotalFinaElf), Esso (later Exxon) and Mobil. To the detriment of Iraq, BP happened to own control of Iran's entire oil production, while Exxon and Mobil grabbed the Saudi Arabian concession. Consequently, due to a conflict of interest between the shareholders, with BP, Exxon and Mobil having more influence and heavier weight, production from Iran, Saudi Arabia and other Arabian Gulf producers, enjoyed a higher level of priority than production from Iraq. Apart from developing the Kirkuk oilfields, exploration and development outside the Kirkuk area was very scarce. As an example, the Ain Zaleh field, which was discovered in 1939, was not developed until 1952 (see Table 3.1).

Table 3.1
Iraqi Oilfield Discoveries up to 1950

Name of Structure	*Discovery Year*	*Reserves (in billion barrels)*
Naft Khaneh	1923	0.30
Kirkuk	1927	22.40
Qaiyarah	1928	1.00
Khashm al-Ahmar	1928	0.15
Khanuqah	1929	0.10
Najmah	1934	1.25
Sadid	1935	0.10
Qasab	1936	0.75
Qalian	1936	0.10
Jawan	1937	1.00
Adaiyah	1938	0.10
Ain Zaleh	1939	0.20
Zubair	1949	4.75
Nahur Umr	1949	6.27
Ratawi	1950	3.00
	Total	**41.47**

Source: *Oil Production Capacity in the Gulf,* Volume 4, Iraq, Book 3 (London: CGES, 1997).

Oil-related activities continued to stagnate in Iraq until 1951 when Iranian oil was nationalized. This nationalization event turned the oil companies' attention to Iraq and a flurry of oil exploration activity ensued, culminating in the discovery of the super-giant South Rumaila and North Rumaila oilfields in 1953 and 1958 respectively, in the Basrah area. As a result, Iraqi oil production, which averaged around 85,000 bpd during 1940–50, exceeded one million bpd by 1961 (see Table 3.2 and Figure 3.2).

Table 3.2
Iraqi Oilfield Discoveries between 1951 and 1961

Name of Structure	*Discovery Year*	*Reserves (billion barrels)*
Butmah	1952	0.10
Bai Hassan	1953	2.50
S. Rumaila & N. Rumaila	1953 & 1958	27.25
Kor Mor	1953	0.10
Jambur	1954	0.50
Gusair	1954	0.10
Atshan	1955	0.10
Alan	1955	0.10
Sarjon	1956	0.10
Makhul	1956	0.10
Rachi	1957	1.00
Ibrahim	1957	0.10
Injana	1958	0.25
West Baghdad	1958	0.10
Tuba	1959	1.00
Samawa	1959	0.10
Pulkhana	1959	0.10
Demir Dagh	1960	0.10
Chemchemal	1960	0.10
Kifl	1960	0.10
Hamrin	1961	1.75
Luhais	1961	1.00
Qara Chauq	1961	0.10
Dujaila	1961	0.10
	Total	**36.85**

Source: *Oil Production Capacity in the Gulf,* Volume 4, Iraq, Book 3 (London: CGES, 1997).

Figure 3.2
Iraqi Oil Production up to 1961
(in 1000 barrels per day)

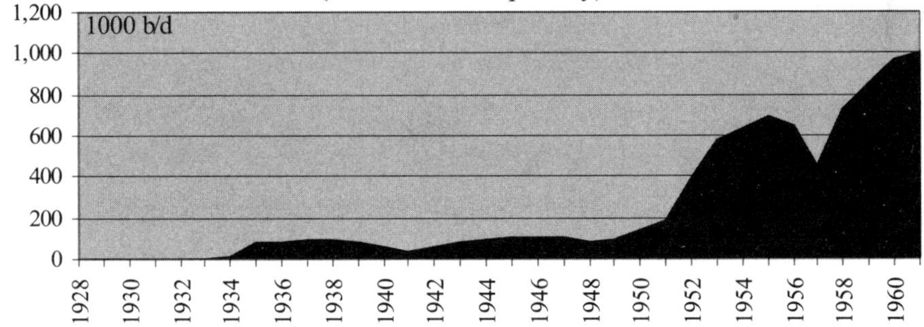

Source: *OPEC Annual Statistical Bulletins,* various issues.

Another Period of Stagnation

After the revolution of 1958, the Iraqi government invited the oil companies for negotiations to address some outstanding issues, including the relinquishment of concession areas left unexplored or undeveloped. Since no agreement was reached with the oil companies, the Iraqi government promulgated Law Number 80 in 1961, expropriating 99.5% of the concession areas, leaving only around 0.05% of the concessions in the hands of the companies—in effect the areas in which they were already operating. This act, which was deemed illegal by the oil companies, ushered in another period of stagnation in oil activity.

From 1961, when Law Number 80 was issued, until 1972 when Iraq began the nationalization process of the three oil companies, Iraq's oil production, which increased from 1007.1 thousand bpd in 1961 to 1465.5 thousand bpd in 1972, did not exceed an average growth rate of 3.4% per year compared with 7.5% for the whole world and 9.7% for OPEC during that same 11-year period (see Figure 3.3). In fact, during 1961–72, Saudi oil production, which started in 1938 (11 years after Iraqi production began), increased from 1480.1 thousand bpd to 6016.3 thousand bpd, making an average growth rate of 12.7% per year, while Kuwaiti oil production, which started in 1946 (19 years after Iraqi production began), increased from 1735.0 thousand bpd to 3283.0 thousand bpd, with an average growth rate of 5.8% per year.[2]

The Decade of Discoveries

With the nationalization of Iraq's upstream oil industry during the first half of the 1970s, and the subsequent settlement of outstanding disputes with the oil companies, Iraq began a period of massive oil and gas activity, which lasted until the beginning of the Iran–Iraq War in 1980. The Iraqi National Oil Company (INOC), which was formed in 1964, started a program of intensive oil exploration with the help of some foreign petroleum service companies. The result of such activity was the discovery, during the 1970s, of important giant fields, including West Qurna (1973), Halfaya (1975), East Baghdad (1976) and Majnoon (1977) (see Table 3.3).

Table 3.3

Iraqi Oilfield Discoveries during the 1970s

Name of Structure	*Discovery Year*	*Reserves (billion barrels)*
Abu Ghirab	1971	2.25
West Quma	1973	8.00
Jabal Fauqi	1974	1.50
Halfayah	1975	3.00
East Baghdad	1976	11.00
Khabbaz	1976	2.00
Majnoon	1977	10.00
Saddam	1978	1.50
Noor	1978	1.00
Nasriya	1979	1.00
Other Fields with <1.0 billion barrels each		3.70
Total		**44.95**

Source: *CGES, 1997: Oil Production Capacity in the Gulf*, Volume 4, Iraq, Book 3, London, UK.

In a matter of nine years, around 45 billion barrels of new reserves were added, and Iraq's total proved reserves exceeded 100 billion barrels by 1980. Iraq's oil production increased dramatically from around 1.5 million bpd in 1972 to around 3.5 million bpd in 1979, attaining an impressive average growth rate of over 12% per year during 1972–79 (see Figure 3.3).

Figure 3.3
Iraqi Oil Production between 1961 and 1981
(in 1000 barrels per day)

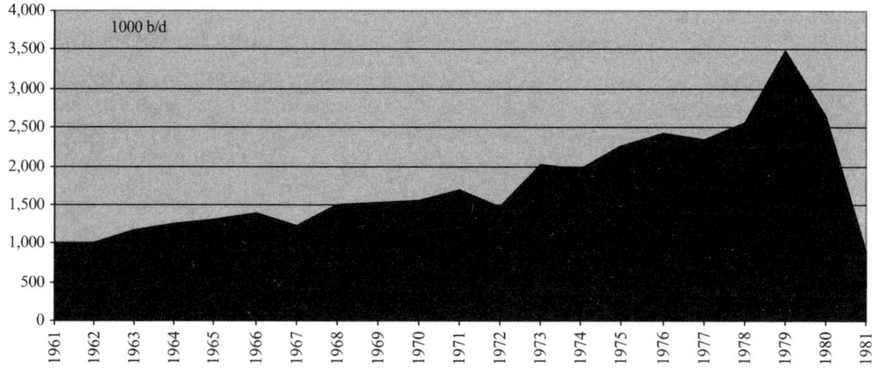

Source: *OPEC Annual Statistical Bulletins,* various issues.

Saddam's Dark Era

However, that incredible expansion was not destined to continue in Iraq after Saddam Hussein assumed full executive power in 1979. The Iran–Iraq War started in September 1980, the same month in which Iraq's oil production peaked at 3.8 million bpd. During the first weeks of the war, Khor al-Amaya, one of Iraq's deep-water terminals in the south, with an export capacity of 1.6 million bpd, was completely destroyed. The other terminal, Mina al-Bakr (later Basra Oil Terminal), also with a capacity of 1.6 million bpd, was partially destroyed. In addition, the Iraq–Syria pipeline was closed in 1982 partly due to the perennial feud between the Ba'ath Party's two wings and partly in a bid by Syria to help Iran.

After the closure of this pipeline, Iraq was left for a time with only 0.7 million bpd export capacity through the Iraq–Turkey pipeline.

As a result of a few years of concerted efforts, however, a parallel line to the Iraq–Turkey pipeline was laid, raising the export capacity through Turkey to 1.7 million bpd, and another pipeline was constructed with a capacity of 1.6 million bpd, linking Iraq's southern oilfields to the Red Sea across Saudi Arabia. The Mina al-Bakr was also repaired, restoring a deep-sea export capacity of 1.6 million bpd. With these expansion and restoration efforts, Iraq's sustainable oil production capacity was raised to 3.5 million bpd and oil production increased gradually from a trough of 897.4 thousand bpd in 1981 to another peak of 2785.8 thousand bpd in 1989 (see Figure 3.4).

Figure 3.4
Iraqi Oil Production during the Period 1980–90
(in 1000 barrels per day)

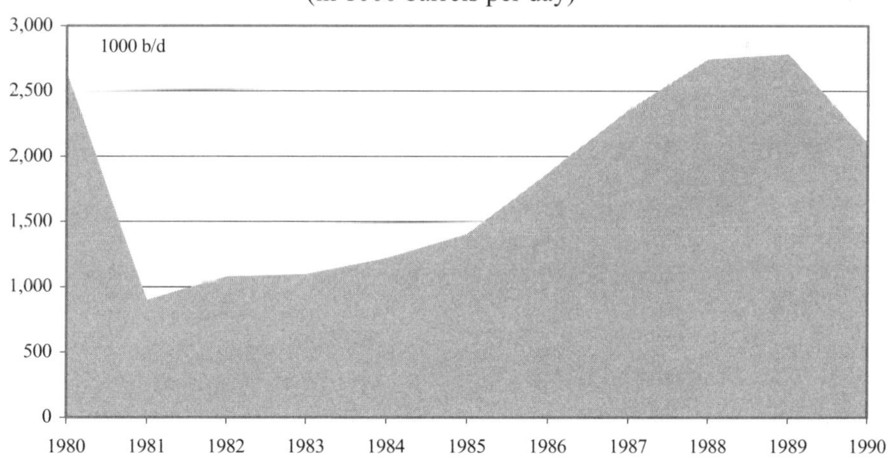

Source: *OPEC Annual Statistical Bulletins*, various issues.

The invasion of Kuwait in 1990 precipitated the destructive Gulf War, followed by a long period of economic sanctions. During this war, the aerial bombardment caused extensive damage to Iraq's oil infrastructure, including producing wells, degassing stations, pumping stations, storage tank farms, processing plants, pipelines and oil export terminals. As a result, Iraq's oil production collapsed from over 2.1 million bpd in 1990 to a mere 286 thousand bpd in 1991 (see Figure 3.5). Whilst Iraq was able to rehabilitate much of its damaged oil infrastructure after the cessation of the Iran–Iraq War, it was unable to do so in the aftermath of the Gulf War because of the imposition of economic sanctions. Iraq was formally prohibited from importing machinery, equipment and supplies for the oil sector, and even if it was able to import such goods, it could not have paid for such imports because of the lack of foreign exchange.

Figure 3.5
Iraqi Oil Production during the Period 1990–2003
(in 1000 barrels per day)

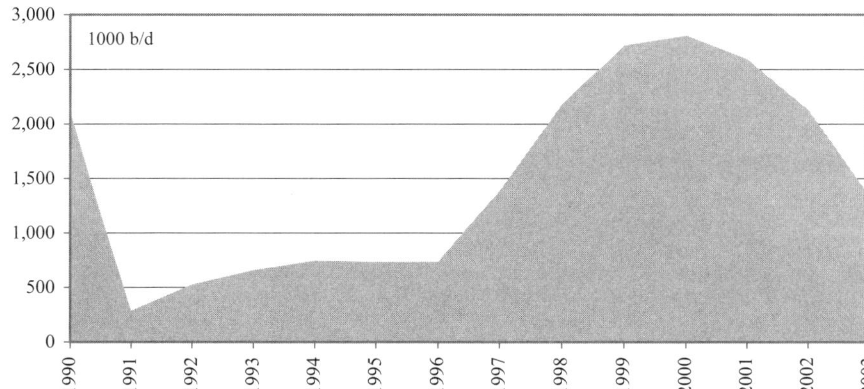

Source: *OPEC Annual Statistical Bulletin* for 1990–2002 and *BP Statistical Review* for 2003.

Under an environment of complete sanctions, Iraqi oil production stagnated since it was confined to meeting domestic demand, supplying Jordan's needs (with the tacit consent of the UN), and some crude and products smuggling, mainly through Turkey and the Gulf. Oil production did not start recovering until December 1996 when the Oil-for-Food Program was finally agreed upon and put in practice. As a consequence of that program, oil production rose rapidly during the four years after 1996 until it reached a peak of 2.8 million bpd in 2000.

Thereafter oil production assumed a declining trend until the regime's collapse in 2003 when, in that year, it averaged only 1.3 million bpd (see Figure 3.5). The main reason for the decline during 2000–2003 was due to differences between the United Nations and Iraq on oil pricing, with Saddam stopping oil exports now and then to achieve political ends. Finally, during the invasion, production from the south ceased with the start of the war on March 20, 2003 and production from both north and south ceased completely when Baghdad fell on April 9, 2003.

Mismanagement and Destruction

With the start of the Oil-for-Food Program, the Iraqi regime followed a policy of maximizing oil production to the extent allowed under the program ceilings, with the aim of maximizing oil revenues. Iraqi oil production notably accelerated after the program ceiling was lifted at the end of 1999 and Iraq was allowed to produce to the extent of its capability, until the gradual erosion of its exports as a result of its disputes with the United Nations. However, accelerated oil production was not conducted in accordance with sound oilfield practices.[3] Proper reservoir management necessitates constant monitoring, including the carrying out of surface and sub-surface measurements of static and dynamic parameters. Such monitoring was either neglected or hampered during the years of sanctions on account of the loss of specialized equipment, lack of modern technology and the absence of service companies. Elevated production under such opaque conditions led to the loss of many producing and injection wells in the north and south, and may have caused permanent loss in the ultimate reserves of the Kirkuk and Rumaila fields. As an indication of the damage sustained by some oilfields, pumping to Ceyhan (on the Mediterranean) via the Iraq–Turkey pipeline was temporarily delayed during March 2004, due to technical problems, despite completing repairs to the sabotaged pipeline. Rising water cut (the damaging intrusion of water into oil reservoirs) was the technical problem suffered by the Kirkuk field.[4]

The invasion of Iraq further damaged the country's oil sector. Damage was inflicted on all Iraqi companies involved in the industry, including the North Oil Company (NOC), South Oil Company (SOC), North Gas Company (NGC), South Gas Company (SGC), the three refining companies in the north, centre and south, Oil Products Distribution Company (SADOP) State Company for Oil Projects (SCOP), Iraqi Drilling Company (IDC) and Oil Exploration Company (OEC). The damage inflicted on the oil industry was wide ranging, from the smallest item to incredibly heavy equipment or large installations. Direct war damage, looting, sabotage or a combination of these were inflicted on degassing stations, oil pumping stations, gas compressor stations, production facilities, industrial water supply and water injection facilities, storage tanks, vehicles, cranes, drilling rigs, field laboratories, office equipment and electric generators. Even a field hospital was looted.

Restoration Plan

In anticipation of potential wartime damage in addition to other possibilities including the occurrence of oil well fires, the US Congress authorized the US Department of Defense (DOD) to carry out tasks such as extinguishing oil fires, carrying out environmental restoration for marine and land-based oil spills, ensuring safe shutdown of oil facilities during initial stages of the war, and assisting the Iraqi Ministry of Oil (IMOO) in restarting and operating the upstream and downstream oil industry after the end of the war. The DOD, in turn, assigned the above-mentioned responsibilities to the US Army Corps of Engineers (USACE). Accordingly, a unified comprehensive plan to restore the damaged oil infrastructure to its pre-war state was developed[5] and a total of 220 projects were identified for restoration (see Table 3.4).

As Table 3.4 shows, NOC and SOC sustained most of the damage (represented as projects in the table) with both companies sustaining 105 incidents out of a total of 220 incidents. The total cost of repair was estimated at $1143.7 million,[6] with NOC and SOC taking the lion's share of 57 percent. Due to the damage inflicted on the upstream oil industry (looting of oilfields and damage to degassing stations), Iraq lost over 1.9 million bpd in production capacity (see Table 3.5). All repairs, according to the plan, were to be completed by April 2004, whereby the pre-war production capacity of 2.8 million bpd was to be restored.

Table 3.4
Distribution of Damage and Cost According to the Restoration Plan

Company	*Number of projects*	*Total estimated cost (million$)*	*% of Total*
NOC	48	320.3	28.0
SOC	57	331.9	29.0
NGC	7	31.9	2.8
SGC	9	23.8	2.1
Refineries	29	42.9	3.8
Downstream	29	148.9	13.0
SCOP	4	43.5	3.8
IDC	26	132.0	11.5
OEC	11	68.5	6.0
Total	**220**	**1143.7**	**100.0**

Source: Task Force Restore Iraqi Oil (RIO), US Army Corps of Engineers (USACE), 2003, Baghdad, Iraq.

Table 3.5
Iraqi Oil Production Capacity Before and After the Invasion
(in 1000 bpd)

Company	*First quarter '03 (before the invasion)*	*May '03 (after the invasion)*	*Capacity loss*
NOC	900	620	280
SOC	1,940	280	1,660
Total	**2,840**	**900**	**1,940**

Source: Iraqi Ministry of Oil.

Failure to Deliver

The target of regaining the desired capacity of 2.8 million bpd by April 2004 was not achieved. Instead, a production capacity of around 2.4 million bpd was reached, and the current capacity seems to continue at this level. A number of factors have prevented the capacity from reaching the target of pre-war level. Among them are:

- The American companies involved claim that the large number of projects to be implemented, frequent additions of newer requirements and modifications, and complex US bureaucratic procedures in getting approval for project funding have contributed to the delay. An example of the important additions is the replacement of the two water injection pipelines connecting the water treatment plant at Garmat Ali and the South Rumaila cluster pumps. The two 48-inch diameter lines failed frequently because of age. Until they are replaced, not enough water can be injected into the fields, thus contributing to lower production.[7] Furthermore, under the chaos of military occupation, the US contracting companies may not have acted in good faith as they would most likely have done had the contracts been signed directly with IMOO and not with USACE, and had there been peace and stability in the country. Some US congressmen even accused the contracting companies, namely Kellogg, Brown and Root (KBR) of waste, abuse and fraud. Although the total funding allocated to the restoration plan was raised to $1.7 billion (and could go up to $2 billion), only $769 million, or 45% of these funds have been earmarked so far, while the portion of work actually completed accounts for only $142 million, no more than 8% of the funds allocated.[8]
- Regarding the prevailing work environment in Iraq, the worsening security situation, including political unrest, terrorism, murders, killing, kidnapping, hostage-taking (and even beheading) have severely hindered the work of the contractors (KBR in the south and Parsons in the north). Shorter and less productive working days have

been causing delays in completing assigned projects, and rising insurance rates and skyrocketing transportation and security costs have been consuming up to 20% or more of the donated US aid for Iraq's reconstruction.

- On the part of the Coalition Provisional Authority (CPA), the restoration plan to be undertaken by KBR and Parsons only concerned above-ground structures and not underground improvements—unless they were incidental to such work. Partly because of this plan and because IMOO wanted to increase Iraq's production capacity to the pre-Gulf War level of 3.5 million bpd (i.e. beyond the pre-invasion capacity of 2.8 million bpd adopted by the restoration plan), IMOO had to follow its own plan to increase the country's capacity to the desired level. This plan entailed comprehensive reservoir and engineering studies for the Kirkuk and North and South Rumaila oilfields, to be followed by pertinent well workovers and infill drilling.[9] However, such a plan was frustrated by severe lack of funding. The IMOO budgets for 2003 and 2004 were the bare minimum for operational needs. The CPA, which was in charge of budgeting for those two years, failed to allocate capital expenditures for the upstream oil sector. It was only during its May 15, 2004 meeting (just about one month before it was disbanded) that the CPA hastily allocated about $2 billion for several projects, of which $460 million was allocated to the IMOO for oil infrastructure rebuilding.[10]
- Several other factors contributed to the delay of the above-mentioned contractors' work, and these included frequent power supply interruption, Iraqi socio-economic problems, inefficient labor and the general weakness of the government. Besides, the sabotage of export pipelines (specifically the Iraq–Turkey pipeline) led to lower oil exports and less oil revenues, while sabotage of crude oil pipelines feeding the domestic refineries led to a serious shortage in oil products, which had to be covered by imports. Additionally, the product import problem was further exacerbated by the incessant

smuggling of oil products to neighboring countries, enticed by the huge price differentials. Consequently, the reduced oil revenues and high products import bill, in addition to the CPA's budgeting short-sightedness, led to the investment-capital starvation of IMOO and contributed to the eventual lack of achievement in capacity restoration.

Iraqi Oil: What Future?

For the short term, the two American companies, KBR and Parsons, continue to work on the oil infrastructure restoration plan to regain the pre-war production capacity of 2.8 million barrels per day. The contracts are funded through US Congress appropriations, which may go up to $2 billion, but the two companies are behind schedule in their work, as described above.

For the medium and long term, the Iraqi Ministry of Oil has announced a plan, which includes increasing Iraq's sustainable oil production capacity to 3–3.5 million bpd by the end of 2005 and to 6 million bpd by 2010.[11] Also, one of the plan's aims is to provide the country's electric power stations with natural gas, to replace the burning of oil for electricity generation. Using natural gas to produce electricity has the advantage of higher efficiency when using combined cycle gas turbines (CCGT), and provides better protection to the environment since natural gas has lower carbon content than the other fossil fuels. Besides, the burning of natural gas, considerable amounts of which are being flared in Iraq at present, displaces higher-rent oil and frees it for export.

Capacity Expansion in the Medium Term

Assuming that KBR and Parsons are successful in implementing the restoration plan and in recovering the pre-invasion capacity of 2.8 million bpd, the issue of increasing Iraq's sustainable production capacity to 3 million bpd or 3.5 million bpd by the end of 2005 should not cause a problem for the IMOO, whether technically or financially. From the technical standpoint, the first priority would be to conduct comprehensive reservoir and engineering studies of the key producing fields, namely the Kirkuk, and the North and South Rumaila oilfields. The last studies of such nature were carried out in the early 1980s for the Kirkuk field and in the mid-to-late 1980s for the two Rumaila fields. After this long period of neglect it has become absolutely vital to conduct the required studies in order to spare further damage to these super-giant fields, and the IMOO has already issued a tender in this respect.[12]

In order to meet the 3.5 million bpd target, additional production will come from the existing producing fields by tapping new pay zones in already drilled wells, and also through increased production from the appraised and partially developed fields. The bulk of oil production at this stage, however, will come from the Kirkuk and Rumaila fields, and the main aim of undertaking the reservoir and engineering studies is to optimize production by means of well workovers and infill drilling so that ultimate recovery from these fields is maximized.

Since most of the infrastructure is already in place in producing fields, adding new capacity from such fields would be less costly than green field development. One rather low estimate, puts the addition of one daily barrel of production capacity at $1900 in the southern fields and only $950 in the northern fields.[13] Another estimate puts such cost at $3,000–4,000 per daily barrel.[14] In order to reach a capacity of 3.5 million bpd, 0.7 million bpd must be added over and above the restored 2.8 million bpd pre-invasion capacity. If we adopt the more credible estimate of $3,000–4,000, from IMOO officials, then the total investment needed would be in the range of $2.1–2.8 billion. The IMOO may not independently be able to handle such expansion within the set time limit (end of 2005). In this case, all or part of the 0.7 million bpd capacity addition may be implemented with the help of oil service companies via service contracts. Budgeting of $2.1–2.8 billion by the Iraqi government for 0.7 million capacity expansion at these times may be difficult, and if no such money is available in the treasury for this kind of investment, financing through borrowing can be used, with oil as collateral. Due to the extremely high rent of oil, expenditures of $2.1–2.8 billion can be recovered very quickly. For example, to pay out a $2.8 billion loan, via additional oil exports of 0.7 million bpd, at a net value of $33.5 per barrel (assuming a price of $35 per barrel less production cost of $1.5 per barrel), only four months of such exports are needed.

In parallel with this operation, the IMOO needs to rehabilitate the long-neglected oil infrastructure in order to handle a capacity of 3.5 million bpd with respect to production, processing, storage, transport and export operations. In this regard IMOO, as a result of recurring war damages, has accumulated considerable experience in the areas of repair, replacement and reconstruction, and it is estimated that a sum of about $1–2 billion is needed for infrastructure repair and refurbishment[15] with the expenditure amount depending on the extent of this operation. As part of this refurbishment operation, and in order to improve the country's crude oil transportation by pipeline, the IMOO has already allocated $250 million to modernize all pipelines carrying crude oil to Iraq's export terminals and also to pipelines carrying fuel to power stations.[16]

Capacity Expansion in the Long Term

For its long-term plan, the IMOO has recently announced that a production capacity of 6 million bpd is planned for 2010. Iraq's recoverable oil reserves can, of course, support this and even higher levels. Iraq's estimated proved recoverable reserves were 115 billion barrels at the start of 2004,[17] and its oil potential (undiscovered recoverable reserves) was estimated by the Centre for Global Energy Studies (CGES) to be around 216 billion barrels.[18] However, the actual level of oil production that can be reached depends on the depletion policy that a country might choose to follow.

For example, based on their oil production rates and published reserves in 2003, the depletion rate (percentage of reserves produced in that year) was around 1.4% for Saudi Arabia, 4.5% for Russia, 8.8% for the USA, 11.8% for Norway and 18.5% for the UK.[19] Given that Iraq was long deprived of the opportunity to export oil due to multiple factors (including wars, sanctions and sabotage) Iraq would surely want to maximize its production and exports. Based on the presently proven recoverable reserve base of 115 billion, a production rate of 6 million bpd represents a depletion rate of 1.9%, a production rate of 8 million bpd represents a depletion rate of 2.5%, while a production rate of 10 million bpd represents a depletion rate of 3.2%, which is even smaller than Russia's depletion rate of 4.5%—although Russia's proved reserves make only 60% of the Iraqi reserves.

If Iraq assumes a depletion rate similar to that of Russia (4.5%), then Iraq's production rate should reach around 14 million bpd. Even such a production level can be sustained for several decades if around 5 billion barrels a year of recoverable reserves is ploughed in from the potential reserves of 216 billion barrels.

Oilfield Development

Due to the unusual conditions that Iraq has experienced, the majority of its discovered and appraised oilfields remain undeveloped. Out of 73 such fields, only 15 have been developed. With the development of such giant fields as Majnoon, West Qurna, Halfaya and Nahr Umar, and many other smaller fields, it would be very easy to add 2.5 million bpd to the pre-Gulf War capacity of 3.5 million bpd to reach the targeted goal of 6 million bpd by 2010, as planned by IMOO.

In a paper presented by high-ranking IMOO officials at a Baghdad seminar,[20] the authors claimed that 450,000 bpd can be developed from 11 appraised fields in the north (Table 3.6A), 3 million bpd can be developed from 11 appraised fields in the south (Table 3.6B) and 300,000 bpd can be developed from 3 appraised fields in the central area (Table 3.6C), making a total of 3.75 million bpd from 25 undeveloped fields in the north, south and centre of Iraq. Add to that another 900,000 bpd, which can be developed from 8 producing fields in the south, and the total would be 4.65 million bpd (Table 3.6D).

Table 3.6A

Production Potential of the Northern Appraised Fields

	Production Rate (1,000 bpd)
Partially Developed Fields:	
Humreen	
Khurmala	
Sub Total	**160**
Light Oil North Eastern Fields:	
Galabat	
Taq Taq	
Qamar	
Qara Choq	
Khashm Al Ahmar	
Sub Total	**120**
Heavy Oil Fields:	
Qayara	
Qasab	
Nejmeh	
Jawan	
Sub Total	**170**
Total (11 fields)	**450**

Source: Iraqi Ministry of Oil, reported in the *Middle East Economic Survey* (MEES) March 20, 1995.

Table 3.6B
Production Potential of the Southern Appraised Fields

	Production Rate (1,000 bpd)
The Four Giant Fields:	
Halfaya	
Nahr Umr	
Majnoon	
West Qurna	
Sub Total	**2,100**
Fields of the Gharaf Axis:	
Gharaf	
Nasiryia	
Rafidain	
Sub Total	**390**
Misan Fields:	
Amara	
Noor	
Sub Total	**100**
West Basrah Fields:	
Tuba	
Ratawi	
Sub Total	**410**
Total (11 fields)	**3000**

Source: Iraqi Ministry of Oil, reported in the *Middle East Economic Survey* (MEES) March 20, 1995.

Table 3.6C
Production Potential of the Central Appraised Fields

	Production Rate (1,000 bpd)
Baghdad Province:	
East Baghdad	
Balad	
Ahdab	
Total (3 Fields)	**300**
Total North, South and Central Appraised Fields (25 Fields)	**3,750**

Source: Iraqi Ministry of Oil, reported in the *Middle East Economic Survey* (MEES) March 20, 1995.

Table 3.6D
Production Potential of Additional Development of Producing Fields

	Production Rate (1,000 bpd)
Additional Development of Producing Fields: Mishrif Reservoir:	
South Rumaila	
North Rumaila	
Zubair	
Sub Total	**470**
Nahr Umr & Zubair Reservoirs:	
South Rumaila	
North Rumaila	
Luhais	
Subba	
Sub Total	**330**
Mishrif & Asmari Reservoirs:	
Abu Ghirab	
Buzurgan	
Fuqa	
Sub Total	**100**
Total Appraised Fields (8 Fields)	**900**
Grand Total (33 Fields)	**4,650**

Source: Iraqi Ministry of Oil, reported in the *Middle East Economic Survey* (MEES) March 20, 1995.

In the same paper referred to above, the authors put forward an estimate of $4,500–5,000 per daily barrel for grassroots development.[21] If 2.5 million bpd is to be added from developing some of the appraised fields, in order to reach a sustainable capacity of 6 million bpd, the estimated cost would be $11.25–12.5 billion. Such cost covers development within the field boundary; transportation, storage and export facilities would require additional cost.

The present Iraqi oil export capacity is as follows: from the south (in the Arabian Gulf) there is 1.4 million bpd from the Basra Oil Terminal and around 0.4 million from the Khor al-Amaya terminal. In order to regain the original export capacity of 1.6 million bpd for each, a crude oil storage depot will have to be built to serve these two terminals. In addition, both of the terminals will have to be rehabilitated, especially Khor al-Amaya, which was completely destroyed during the 1991 Gulf War. Besides, the problem of Khor al-Amaya's draft has to be addressed since, currently, the maximum draft for a vessel at this terminal is only 17 meters and cargoes cannot exceed 1 million barrels per vessel loading from this terminal.

Another export outlet is in the north, via the Iraq–Turkey pipeline to the Mediterranean port of Ceyhan. Notwithstanding recurrent sabotage to this line, the present capacity of around 0.9 million bpd can be raised to the line's original capacity of 1.7 million bpd if the destroyed IT2 pumping station and a metering station near the border are rebuilt or rehabilitated. Once the export capacities of the two deep water terminals and the Iraq–Turkey pipelines are recouped, Iraq's export capacity will reach 4.9 million bpd, which is more than enough to handle the first stage capacity expansion to 3.5 million bpd.

In order to handle a capacity expansion to 6 million bpd, Iraq will need to persuade Saudi Arabia to return to it the expropriated line with a capacity of 1.6 million bpd extending from the southern Iraqi city of Basra, going across the Kingdom of Saudi Arabia to the Red Sea. Repair of the K3 pumping station in Haditha (currently part of the USACE restoration plan) will allow Iraq to export another 300,000 bpd across Syria to the Mediterranean port of Banias. The original export capacity of the Iraq–Syria pipeline was around one million bpd. Although this capacity has deteriorated over time to the present level of 200,000–300,000 bpd, it can be raised to 600,000 bpd with some maintenance.[22]

To regain the flexibility of exporting Iraq's northern crude through the southern export outlets and the southern crude through the Mediterranean export outlets, the Strategic Pipeline and its pumping stations will have to be refurbished and perhaps, another parallel line would have to be added.

Financing Capacity Expansion

The Iraqi National Oil Company (INOC) was created in 1964 and did a good job in the areas of exploration and capacity expansion, using indigenous talent as well as specialized foreign expertise. INOC continued its work until 1987 when, as a result of Saddam's misrule, it was dismantled into smaller companies to be controlled by the bureaucracy of the Iraqi Ministry of Oil.

In order to revive Iraq's oil industry, INOC should be re-established as an autonomous oil company, with the purpose of running, more efficiently, the daily operations of the oil industry while, on the other hand, allowing the Ministry of Oil to concentrate on planning and policy making, interacting with other Iraqi ministries and agencies, and taking care of OPEC affairs. The present interim Iraqi government has realized the importance of oil industry decentralization and already announced the reconstitution of INOC as well as establishing a Supreme Petroleum Council.[23]

Based on the estimates shown above, Iraq needs about $15–17 billion in order to increase its production capacity from the pre-invasion level of 2.8 million bpd to the pre-Gulf War level of 3.5 million bpd by the end of 2005, and then move to meet the new IMOO target of 6 million bpd by 2010. However, this huge amount is not needed as a single lump sum. It is needed in stages as the capacity itself is increased in phases. With more production capacity being added, more oil revenue is generated and the financial capability of the country will be better positioned to address further finance needs.

Over a period of six years (2004–2010) an average of $2.5 billion per year is needed, assuming a total of $15 billion will be sufficient for the capacity to reach $6 million bpd. However, given Iraq's dire need for money to meet the government's ordinary or current budget expenditures, to service foreign debts, to pay for war reparations and to finance the process of reconstruction, the Iraqi government may not be financially capable of budgeting such amounts to fund the envisaged capacity expansion. In such a case, INOC can finance this operation without any recourse to the Iraqi treasury.

Among several options, there are three obvious ways for the financing of the envisaged capacity expansion: first, INOC could conduct successive borrowing transactions; second, INOC could invite international oil companies (IOCs) to carry out the needed expansions, using the IOCs' money and technology; and third, a combination of the first two could be used. A fourth possible way could be to consider privatization, an option that will be discussed later.

Going it Alone

It is true that Iraq, as aforementioned, is presently in a dire financial situation and cannot afford to allocate billions of dollars for oil industry rehabilitation and oilfield development. Nevertheless, INOC, as an autonomous oil company with billions of proved oil reserves, can tap the capital markets to finance the desired capacity expansions. Indeed, lenders would be very happy to arrange for the required financing, since only rehabilitation and refurbishment of the producing fields and associated infrastructure is needed to reach a target of 3.5 million bpd, and only the development of already discovered and appraised fields is needed to reach 6 million bpd. In the Iraqi case, there are several advantages. Exploration is not necessary and hence there is no exploration risk involved. Iraq's oil is now the cheapest in the world in terms of the costs needed for development and production. Moreover, Iraqi oil is close to two major consuming markets—Europe through the Mediterranean outlets and Asia through the Arabian Gulf outlets.

The majority of Iraq's oil reserves are contained in super-giant and giant fields, as Table 3.7 shows. Out of 15 developed fields, which contain less than 40% of the country's total proved reserves, 68.7% of the reserves are housed in just the top 3 fields, and over 92% of the reserves are housed in the top 8 fields. Out of the 58 appraised but undeveloped fields, which contain over 60% of the country's total proved reserves, 60.4% of the reserves are housed in just the top 4 fields and over 86% of the reserves are housed in the top 12 fields.[24]

Table 3.7

Distribution of Iraq's Developed and Undeveloped Reserves

(in million barrels)

	More than 5000	*1000–5000*	*500–1000*	*Less than 500*	***Total***
Developed fields:					
Number of fields	3	5	4	3	**15**
Share of developed reserves (%)	68.7	23.4	6.7	1.2	**100**
Undeveloped fields:					
Number of fields	4	8	7	39	**58**
Share of undeveloped reserves (%)	60.4	26.0	6.9	6.7	**100**

Source: Adapted from Ghazi M. Haider, "Economics of Oil Fields Development Ventures in Iraq," in *Middle East Economic Survey* (*MEES*), February 23, 2004.

Other than the East Baghdad field, which is structurally more complex, most of the fields to be developed are located in the south and consist of simple anticlines, thus reducing to a large extent the chance of drilling dry holes. Furthermore, Iraq's producing and prospective reservoirs are generally, not very deep compared with worldwide reservoir depths. However, the southern reservoirs, with an average depth of 2,000–4,000 meters, are deeper than those of the north, where some reservoir depths are even less than 1,000 meters. All these characteristics, in addition to simple terrain and cheap labor, contribute to lower development costs.[25]

High Rent

Amid the current high world demand for oil and the uncertainty of stable supplies from Saudi Arabia, Iraq, Nigeria, Venezuela and lately Russia, the waning spare production capacity of the world has declined to the alarmingly low level of around 1.3–1.5 million bpd (almost all of it based in Saudi Arabia). In their frantic efforts to offset the annual natural decline of 2% and to add a little bit of new capacity, the Saudis are bringing into production the four fields of Qatif, Abu Sa'fa, Manifa and Khurais. Bernstein Investment Research, in a recent report, estimated that the cost of developing these fields would total $19.5 billion, so as to add a further production of 2.4 million bpd by 2012.[26]

This estimate means that developing one daily barrel of production from these four fields costs over $8,000, given that oilfield development in Saudi Arabia is among the cheapest in the world. Yet, oilfield development in Iraq is estimated to cost around $4,500–5,000 per daily barrel as previously mentioned. This is, indeed, the cheapest in the world!

To estimate the potential rent per barrel of, for example, the Majnoon field in southern Iraq, the development of a 0.6 million bpd production capacity from this field would cost $3 billion (taking the higher figure of $5,000 development cost per daily barrel). If INOC borrows the $3 billion amount to be repaid over 20 years at the rate of 10% interest compounded annually, the debt service (principal and interest) would be around $352 million per year, or around $1.6 per daily barrel. This is for capital cost, and if $1.5 per barrel is added to it for production cost, the total cost per barrel would be $3.1. Adding to that an average transportation cost to the export terminal of around $0.4 per barrel, then the total FOB cost to INOC would be $3.5 per barrel. If this oil is sold at $35 per barrel, the rent to

INOC would be a sheer $31.5 per barrel. Who would not be willing to lend INOC the development cost of such high-rent oil, with plentiful reserves acting as debt collateral?

Sharing with the IOCs?

The other option of financing capacity expansion in Iraq is to have international oil companies assume such an undertaking, using their money and state-of-the-art technology. The most popular contractual regime, from the standpoint of the IOCs, for the development of the appraised but undeveloped fields would be production-sharing agreements (PSAs). In the case of Iraq, the previous Ba'ath regime, before its collapse, entered into negotiations and signed agreements in 1997 with such countries as China for the development of the Ahdab field, and Russia for the development of West Qurna, which was later cancelled by the same regime for non-compliance. Another two agreements were negotiated with the French company TotalFinaElf for the development of Majnoon and Nahr Umar fields, but were never signed.

These PSAs, termed by Iraq as development and production contracts (DPCs) for the development of discovered fields, stipulated, in general, that the contractor group (including an Iraqi entity with up to 25% working interest) was to fund the cost of field development, which cost was to be recovered from 40% of oil production, known as "cost recovery oil." In addition, the contractor was to obtain a share of the remaining oil, known as "profit oil," the share of which was to range between 9.7% and 12% of profit oil. The term of the DPC was for a period of up to 23 years.[27]

In 1998 Iraq adopted a modification to the DPC called "Modified Buy-Back," which was less popular from the IOCs' standpoint. Such contracts stipulated a term of 12 years comprising three phases: a 5 to 6-year development phase, a 2 to 3-year transitional phase (for gradual handover to an Iraqi entity), and a final phase covering sustained production. The contractor would be the operator during the first two phases, being under the supervision of a Joint Management Committee. The Iraqi entity would share the contractor with a 10% working interest (i.e. sharing 10% of the cost and receiving 10% of the interest).

Under such a contract, the contractor group (including the 10% owning Iraqi entity) would finance the development costs. The whole contractor costs, which include development and operating costs and any bank loan expenses, would be recouped from up to 50% of production revenue. In addition, the contractor would be paid fees to cover an agreed rate of return (ROR) on the invested capital, with such ROR being subject to a floor and ceiling governed by a certain percentage of the capital invested, and would be paid from up to 10% of production revenue. Unlike PSAs, where the contractor receives his oil entitlement at the loading end, the Modified Buy-Back agreement does not give the contractor rights to production but may include an option for lifting, by the contractor, of up to 25% of the field's production. Also, while PSAs give the contractor a certain percentage of "profit oil" throughout the term duration, the contractor, under the Modified Buy-Back, would no longer have an interest in the field once the costs and agreed upon remuneration is received. However, a 15-year technical assistance agreement could be triggered once the handover is implemented.

Why Share the Rent?

As argued above, Iraqi oil enjoys the highest rent in the world by virtue of the certainty of its presence, as well as low development and production costs. This rent, of course, belongs to the Iraqi people and should be enjoyed by them. In addition to its high rent, Iraq's oil export terminals are located near two major markets (Europe and Asia) and also near deep waters for export to the United States. To raise Iraq's oil production capacity to 6 million bpd should be easy both technically and financially, and to do that Iraq can select one or more versions from an array of contracts tailored to its needs, ranging from service contracts to buy-back and production sharing agreements.

Service contracts would be most suitable in the case of adding around one million bpd from within the presently producing and partially developed fields, since this kind of project does not need more than making technical assessments of the producing oil fields, drilling the required number of injection and producing wells (in this case mostly infill) and carrying out the necessary well workovers and re-completions. On the other hand, developing the appraised but undeveloped fields will require more sophisticated technology and up-to-date management expertise and definitely more money. Here, the help of international oil companies may be enlisted through production sharing agreements.

Then again, why should Iraq share the benefits of its high-rent oil with IOCs if it can (through INOC for example) get the money from lending institutions and obtain modern, cutting edge, technology via outsourcing? As professor Helmut Merklein points out, this is what IOCs normally do. [28]

Furthermore, developing Iraq's super-giant oil fields using PSAs would seem to be detrimental to exercising control over its oil output as long as a considerable portion of its production remains under PSA control. This would be particularly crucial when, sooner or later, Iraq becomes bound by OPEC quotas, in the same way as the other 10 members are bound now.[29] This would be the case when OPEC, in the future, acts to drive the price of its basket of crudes to remain within a desired range, by imposing a lower production ceiling and in the process, requiring Iraq to cut back its production. Under such circumstances, and in the presence of considerable production under PSAs, the Iraqi entity in charge of the upstream oil sector (be it INOC) would have to curtail its own production to comply with the smaller OPEC quota, since foreign companies participating in these PSAs would find it detrimental to their interests to have their production subject to OPEC quotas, and would rather expect to see INOC's production curtailed.

The Vital Pre-condition

To enter into contractual agreements for oilfield development – taking Iraq's present conditions into consideration – the IOCs would naturally prefer dealing with a legitimate Iraqi government, preferably constitutionally elected. More important still, any foreign entity operating in Iraq would want a vital pre-condition to be satisfied, and that is the prevalence of a stable domestic environment with assured security of life and property. The present security situation in Iraq is not safe, either for people or for property.

The complete lack of security in Iraq in the aftermath of the war may be gauged from the following example. During the 16 months between June 12, 2003 and September 12, 2004, 118 attacks on oil installations and personnel in the north, centre and south of Iraq have taken place. These are only the recorded attacks as documented by *Iraq Pipeline Watch*[30] (see Table 3.8). It is probable that scores of other attacks on the country's oil infrastructure during this period went undocumented. Lack of security is not confined to the oil sector, which in fact, is more protected than other sectors of the economy. People from all walks of life in Iraq are affected by the severe lack of security, with robberies, street crimes, car-jacking, kidnapping, terrorist attacks and suicide bombing being rampant. These acts have dealt a fatal blow, for the time being, to the prospect of foreign companies investing or working in Iraq.

In addition to losses in human life and property, the attacks on the oil sector have led to huge losses in crude oil exports and supply of domestic refined products. The Iraqi Oil Minister has stated[31] that since the downfall of Saddam, acts of sabotage to the pipelines for crude oil exports have caused a loss of $2.7 billion in oil revenues, and the attacks on the pipeline grid have led to an additional loss of $3 billion as a result of recurrent losses in domestic refined output, necessitating huge imports of gasoline and other products.

Table 3.8

Documented Attacks on Iraqi Pipelines, Oil Installations and Oil Personnel

(16-month period spanning June 12, 2003–September 12, 2004)

Between 12 June, 2003 and 12 September, 2004, 118 documented attacks on oil installations in the north, center and south of Iraq have taken place. These attacks can be divided as follows:

- Two attacks on oil wells, one producing and one non-producing (both incidents causing fire).
- Nine on storage tank farms and their lines.
- Eighteen on the Iraq–Turkey crude oil export pipeline.
- Twelve on crude oil pipelines feeding the Port of Fao and the two Basra deep-water terminals (Basra Oil Terminal and Khor al-Amaya Terminal).
- One (failed) suicide attack on the Basra Oil Terminal.
- Thirteen on gas and crude pipelines feeding power stations.
- Eight on oil products pipelines.
- Eight on the Strategic Pipeline.
- Sixteen on crude oil lines feeding the Bayji refinery complex.
- Seventeen on crude oil lines feeding Daura refinery.
- Four on crude oil lines feeding southern refineries.
- Nine attacks on oil personnel (total of 11 killed and 17 wounded).
- One attack on South Oil Company headquarters and warehouses.

Source: *Iraq Pipeline Watch*, "Attacks on Iraqi Pipelines, Oil Installations and Oil Personnel" (www.iags.org/iraqpipelinewatch.htm).

While working in Iraq as Oil Advisor (during May–August, 2003) I obtained from the Ministry of Oil a chart showing the incidents of crude charge disruption to Daura refinery, which is located in the Baghdad area. Out of 13 disruption incidents, 9 were due to pipeline sabotage, which took place in a span of less than four months (see Figure 3.6). In addition to the chaos and public discontent caused by the shortage of refined products, the billions of dollars spent on oil product imports, which continue today, have exacerbated Iraq's financial impoverishment and hindered the rehabilitation process of the oil sector.

Figure 3.6
Crude Charge Disruption to Daura Refinery
(April 18–August 15, 2003)

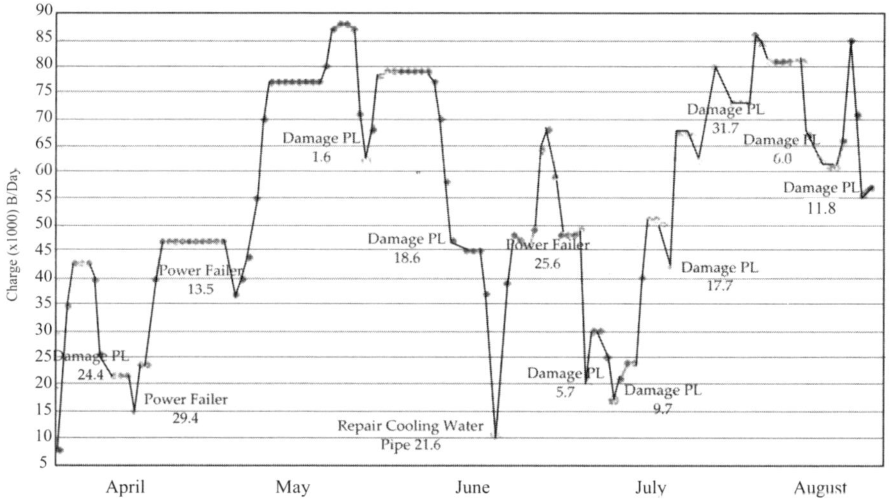

Source: Iraqi Ministry of Oil.

Accommodating Iraq

Although Iraq is a member of OPEC, this organization does not presently bind Iraq by its quota system. In consideration of Iraq's special circumstances, OPEC has kept Iraq out of the quota system since the latter resumed oil exports in accordance with the Oil-for-Food Program, which became effective in December 1996. This special treatment continues currently and may continue for the foreseeable future, despite the downfall of the Iraqi regime in April 2003 and the Oil-for-Food Program ending seven months later.

The key factor presently hindering Iraq from increasing its production is its dire security situation. Once domestic security improves, Iraq will certainly recover its pre-invasion capacity of 2.8 million bpd, and later its pre-Gulf War capacity of 3.5 million. As conditions in Iraq gradually return to normal and Iraqi doors open and conditions become safe for investors, credit institutions and contractors of all kinds, the prospect of the country reaching a sustainable production capacity of 6 million bpd by 2010 will become very real. With such a potential rise in Iraqi production in the foreseeable future, two legitimate questions will arise:

- Can the world oil market accommodate Iraq's new, and rather large, oil supplies?
- With the expected rise in Iraqi oil production, will OPEC continue to give Iraq special treatment and let it produce freely without being quota-bound? Or will it include Iraq within its quota system and hence constrain its production?

To answer the first question, one needs to look at Table 3.9 showing a selection of authoritative forecasts of world demand for oil. All these forecasts seem to agree that world oil demand will continue to increase within the forecast periods. The disagreement, however, concerns the forecasted demand levels and growth rates. According to these forecasts, world oil demand in 2010 will range from a low of 87 million bpd to a high of 91 million bpd, and in 2020 it will range from a low of 100 million bpd to a high of 110 million bpd, with annual growth rates over the period 2000–2020 ranging from 1.4% to 1.8%.[32] Given that non-OPEC supply is not expected to witness any considerable growth during these periods, there seems to be plenty of room to accommodate potential Iraqi oil supplies. As of the third quarter of 2004, almost all forecasts showed world demand averaging around 82 million bpd for that year, with a total increment of 6 million bpd over year 2000, and an average annual growth rate of 1.9% during this 4-year period. Indeed, demand growth for this year alone will very likely average over 3% and such robust growth was expected to continue during 2005, albeit at a lower rate.

Table 3.9
Comparison of Selected Global Oil Demand Forecasts

	2000	*2010*	*2000–2010*	*2020*	*2000–2020*	*2000–2020*
	Actual	*Forecast*	*Difference*	*Forecast*	*Difference*	*Annual growth*
	(million bpd)	*(million bpd)*	*(million bpd)*	*(million bpd)*	*(million bpd)*	*(%)*
OPEC	76	90	14	107	31	1.7
IEA	76	90	14	107	31	1.7
EIA	76	91	15	110	34	1.8
CGES	76	87	11	100	24	1.4

Source: CGES, OPEC, IEA's *World Energy Outlook 2004*, EIA's *International Energy Outlook 2004*.

As for OPEC continuing to accommodate Iraqi production even if the country succeeds in implementing its envisaged oil development plan, there should not be any problem with that if the pre-sanctions quota shares were to be followed. According to the agreed-upon quota distribution before Iraq's invasion of Kuwait, Iraqi production was to be on par with that of Iran, or 14.5% of the OPEC ceiling. At this rate and a hypothetical OPEC ceiling of 28 million bpd, Iraqi production could go up to 4 million bpd and still remain within the allowed quota. Given Iraq's present political and economic circumstances, a production level of 4 million bpd may not be reached before the end of 2007. Even if Iraq's production increases to 6 million by 2010 and it exceeds any allocated OPEC quotas, the organization, in my view, will look the other way, in recognition of the huge losses that Iraq endured during the wars and sanctions since Saddam came to power, and in consideration of the country's continuing financial plight after the downfall of his regime.[33]

Iraqi Oil: The Pitfall

Iraq obtained, during the past half century, tens of billions of dollars in oil revenues, and is destined to obtain much more in the future if the situation calms down and a credible and functioning government comes to power. However, the real aim of the oil exports is not so much the maximization of the revenues as allowing Iraq to benefit from such income. Since the early 1950s and up to 1980, successive Iraqi governments spent a good portion of the country's oil revenues on building physical infrastructure, and on health and education—which was of great benefit to Iraq. On the other hand, those huge oil revenues had their adverse effects on government attitudes toward the crucial issue of Iraq's governance. The country's ample oil revenues negated the need for public taxation, which, in turn, negated the government's need for obtaining public consent and satisfaction. The result was that Iraq fell under successive governments that were completely indifferent to public aspirations and, at the same time, devoid of any measure of public accountability. Not only did those governments follow economic policies harmful to the nation, but the chaotic political situation also ended with a tyrant hijacking the whole country in 1979, and ultimately, destroying everything that had been built with the country's oil revenues.[34]

The adverse effects of the presence of huge mineral wealth – oil in the case of Iraq – on a country's political and economic situation were recently highlighted by Birdsall and Subramanian.[35] These adverse effects, described as the "resource curse," can be recognized in resource-rich countries, which often exhibit symptoms of having underdeveloped or weak political and economic institutions and sometimes having none at all.[36] Development economists have put forward some explanations for the "resource curse" phenomenon. One of them is the price volatility of the resource – specifically oil – and the disruptive effect of fluctuating income on government expenditures. Another explanation is the over-valued currency of resource-rich countries, dubbed in economic literature as the "Dutch Disease," and its impediment to the growth of export sectors, especially agriculture and manufacturing.

The most significant explanation for the resource curse, however, is believed to be the adverse effect of the high rent of natural resources, glaringly apparent in oil and gas-rich countries. Briefly stated, governments of such countries become self-sufficient in terms of income generated from high oil revenues. This self-sufficiency becomes corruptive in two main ways. First, the government's lack of need for tax-generated income reduces its incentive to promote a diversified tax-based economy strongly underpinned by the private sector. Second, the under-taxed, and mostly untaxed, populace loses the desire to hold government accountable for its policies and actions. The result is weak civil society characterized by the lack of checks and balances and proper growth-promoting institutions.[37] Norway is an exception to this thesis. It is claimed that at the time when Norway became hydrocarbon-rich (mid-1970s), democracy in the country was well-entrenched, along with its relevant institutions, and that situation prevented corruption.

A look at Figures 3.7 and 3.8 helps to illustrate these contentions. Figure 3.7 shows the degree of dependence of OPEC's Arab member countries on petroleum exports, and Figure 3.8 shows the contribution of such exports to the GDPs of these countries; Norway is shown here for comparison. If Iraq is ignored since it experienced war and occupation during 2003, Libya seems to be the most dependent on oil revenues (see Figure 3.7), while the UAE seems to be the least dependent, surprisingly beating even Norway. With regard to diversification of the economy (see Figure 3.8), Libya seems to be the country with the least diversified economy while Algeria has the most diversified one, but coming far behind Norway. Here again, the UAE's economic performance seems to be astonishing since it does not lag far behind Algeria on the one hand, Moreover, thanks partly to the UAE's vibrant tourism and manufacturing sectors, the rate of its economic diversification is the fastest of the group and will surpass Algeria within a short period of time.

If democratic institutions shield resource-rich countries from corruption and stagnation, as in Norway (and Botswana for that matter), this reason does not explain why the UAE fared the best among OPEC's Arab members in the first example and is fast moving to best place in the second example. These countries are only moving gradually towards wider public participation in government, albeit at different speeds. In my view, the reason lies in enlightened governance as witnessed in the UAE.

If the governance of an oil-rich country is more enlightened, oil will become more of a blessing, and less of a curse.

Figure 3.7
Share of Petroleum Exports in Total Exports during 2003 (%)

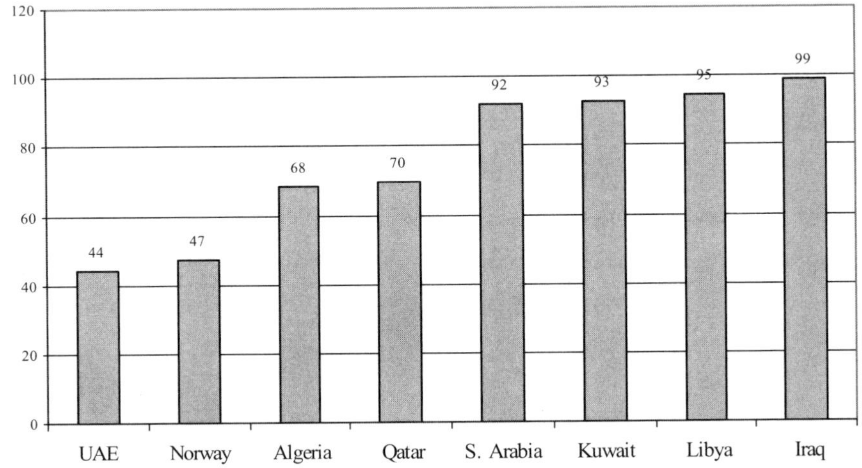

Sources: *OPEC Annual Statistical Bulletin 2003* for Arab countries; and Norwegian Ministry of Finance, Statistics Norway (www.ssb.no/nr_en) for Norway.

Figure 3.8
Share of Petroleum Exports in GDP during 2003 (%)

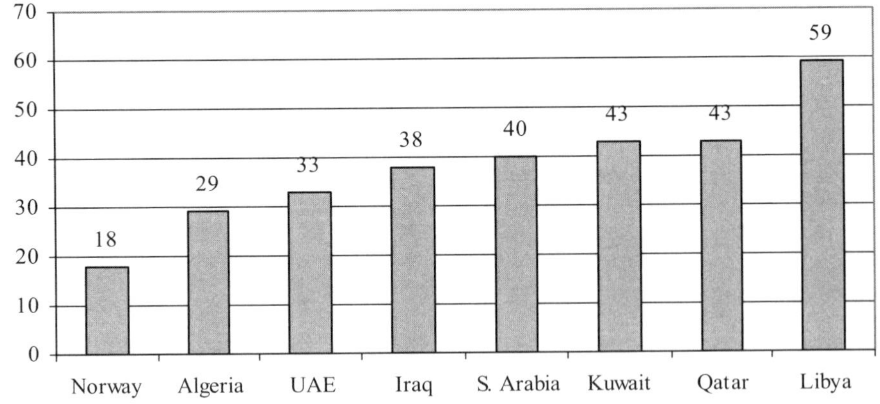

Sources: *OPEC Annual Statistical Bulletin 2003* for Arab countries; and Norwegian Ministry of Finance, Statistics Norway (www.ssb.no/nr_en) for Norway.

The Options Ahead

The potential for abundant natural resources to breed corruption and corrosion in undemocratic countries was glaringly exemplified in Iraq. In the wake of speedily rising oil revenues during the second half of the 1970s, the inherently authoritarian Ba'athist regime was rapidly transformed, with Saddam's absolute monopoly of power, into a system of thuggery in its worst form. In view of this experience, the crucial question becomes: What should be done with Iraq's enormous potential oil wealth?

Several scenarios have been put forward to address the Iraqi situation. One is to privatize the country's oil industry. While it is quite feasible to privatize downstream operations (refining, distribution, etc.), and even upstream-related activities such as drilling and oil service activities, where the rate of return (ROR) on invested capital is normal by virtue of absence of resource rent, upstream oil operations, especially development and production operations, should not be privatized since all the rent resides in that side of the oil resource exploitation. A national Iraqi oil company, run efficiently by a qualified cadre can capture all the rent that ought naturally to belong to the Iraqi people.

Furthermore, adverse consequences can ensue from such privatization. Notwithstanding the concept of sovereignty violation – as the dominant Middle East culture is that resource ownership is a symbol and expression of a country's sovereignty – privatization will lead to accelerated production, since the private sector's discount rate is higher than that of the government. The consequential loss of government control over production will directly conflict with OPEC's role of regulating the market, given that Iraq is a founder member of this organization.

What would Iraq do with the sudden influx of potentially huge oil wealth arising from privatization? It is true that Iraq will need tens of billions of dollars for reconstruction and development. However, the question is: Should the necessary finance for such activities, either from donor countries or from oil revenues, come gradually as needed, because Iraq's absorptive capacity for investment capital is limited? Indeed, one of the top priorities now is to win forgiveness of Iraq's foreign debts to the maximum extent possible. Without such an effort preceding the sale of some Iraqi oilfields, foreign creditors will be swarming around the suddenly cash-rich Iraq demanding payment of their mostly odious debt. The same rationale applies to Iraq's war reparations. Unnecessarily accelerated oil production, which will surely occur under privatization, will lead to accelerated reparation payments, at a time when justice – let alone Iraq's battered economy – requires that the biggest reparation beneficiaries, Kuwait and Saudi Arabia, ought not to punish the Iraqi people for crimes committed by a wildly irresponsible dictator. Indeed, political wisdom dictates that, in order for peace and trust to prevail in a region, countries should act to enrich their neighbors and not impoverish them.

Another proposed scenario is to create an oil fund to even out fluctuating revenues emanating from oil price volatility, thereby stabilizing the government's income. Although this is a good proposition, which is practiced in several countries, it is not a panacea. Such funds can be tampered with by governments and also, the creation of these funds does not shield a country's oil revenues from being misused.

Another interesting scenario is the direct distribution of a country's oil revenues to its people. This has the immediate effect of increasing the per capita income, but may cause inflation, especially if such income is largely spent on consumption. Currently, Iraq's productive base is decidedly too weak and inadequate to cater for the concomitant increase in demand. In case the distributed revenues are allocated largely towards investment, the private sector is much more efficient than government in assessing the markets' needs and in managing the relevant investments. However, it remains true that the private sector only invests in profitable enterprises and stays away from non-profit making endeavors such as spending on infrastructure, health and education. As such, it would be inefficient to distribute the revenues, and then impose taxes on the people in order to finance such activities when, in reality, the government can appropriate from the oil revenues the necessary funds to cover all activities of general public benefit before making any commitment on revenue distribution.

There could be other viable scenarios for efficiently exploiting a country's oil revenues. The important thing, however, is that all of these scenarios are not mutually exclusive. Any combination of these can be adopted to suit the country's unique circumstances. In the case of oil-rich Iraq, the top priority would be to create a diversified, tax-based economy, with little dependence on oil. Nevertheless, so long as there are oil revenues, they should be exploited prudently for the benefit of the country and its people. Given what Iraq has been through during the past two and half decades, the oil revenues should first be allocated to rebuilding the country's infrastructure and for health, education and cleaning the environment. In parallel with that, it would be a good idea to create a modest oil fund for macroeconomic stabilization. Once that is implemented, it would be in the interest of the Iraqi people, the true owners of the hydrocarbon resource, if the government, with the consent and supervision of the legislature, starts making an annual distribution of a reasonable portion of the oil revenues directly to the public. In the meantime, Iraq should march relentlessly towards full democracy. The evolution of a vibrant civil society with its checks and balances, and above all a free press, is a guarantee against potential government corruption and mismanagement of the country's wealth. This system worked for Norway and it should be possible to make it work for Iraq as well.

4

Saudi Arabia's Oil and Gas Industry: Strategic Outlook and Policy Options

Vera de Ladoucette

Grouping oil and gas issues under the same umbrella may seem an obvious categorization. However, such grouping can be misleading as the policy drivers for these two industries are not the same and hence the responses are dissimilar. Moreover, the oil and gas sectors confront different challenges. The only similarity is that, in both cases, the debate has been skewed by false assumptions—the Saudi oil issue has nothing to do with supposedly dwindling reserves and the development of the gas sector is not the remedy to Saudi employment problems.

Saudi Arabia's Oil Industry

At the outset, I would argue that one should not be side-tracked by the recent debate on Saudi oil reserves. Recent concerns about current high prices and low spare capacity have rekindled the debate surrounding Saudi Arabian reserves and the country's long-term production capability. I will not go into the debate about Peak Oil.[1] Suffice it to say that the Cambridge Energy Research Associates in its 2005 report *CERA Worldwide Liquids Capacity Outlook to 2010: Tight Supply or Excess of Riches* sees no reason for concern about oil supply in the medium term. According to the authors Peter M. Jackson and Robert W. Esser, the point at which oil production will peak is not around the corner, but rather a date that looms on the far horizon, sometime after 2020.

Saudi Arabia shows huge potential, which includes possible undiscovered reserves that dwarf other "prolific" areas. There has been a limited exploration effort so far and it is very likely that there are still some giant fields waiting to be discovered.

The proven reserves of Saudi Arabia are reported to be 263 billion barrels, with another 103 billion barrels of probable plus possible reserves according to Saudi Aramco. Saudi Arabia holds 21 percent of world proven reserves and provides some 12 percent of daily production. There are some 85 fields (see Figure 4.1), including a large number of giant fields (see Figure 4.2). CERA sources indicate that only seven of these giant fields account for the bulk of the current capacity—by providing 7.5 million barrels per day (mbpd) out of the 11mbpd production capacity.

The much-publicized doubts about field performance appear to be unfounded, since they are based on limited technical publications that generally focus on production problems and not their solutions.

Figure 4.1
Saudi Arabia's Oil and Gas Infrastructure

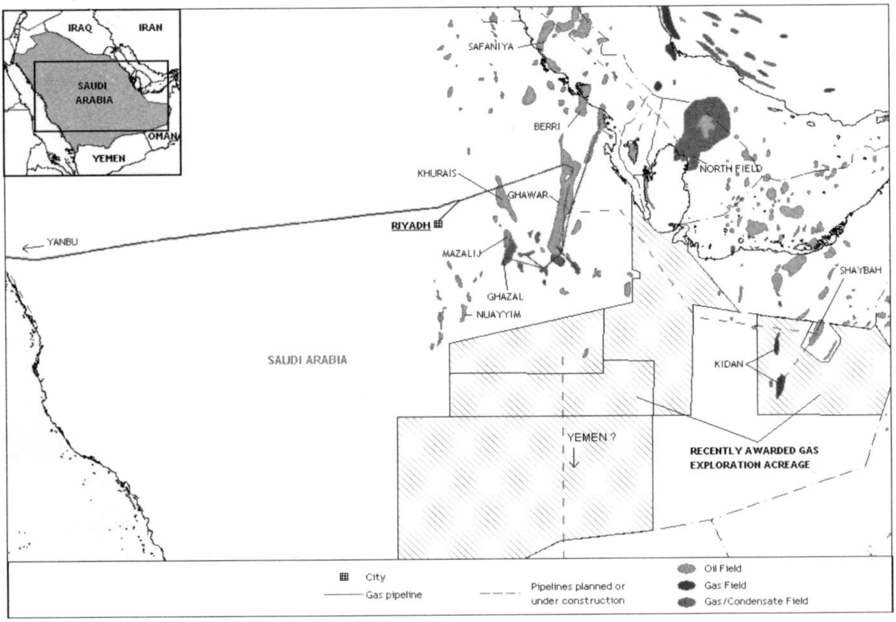

Source: IHS Energy.

Figure 4.2
Saudi Arabia's Giant Field Size Distribution
(excluding Ghawar)

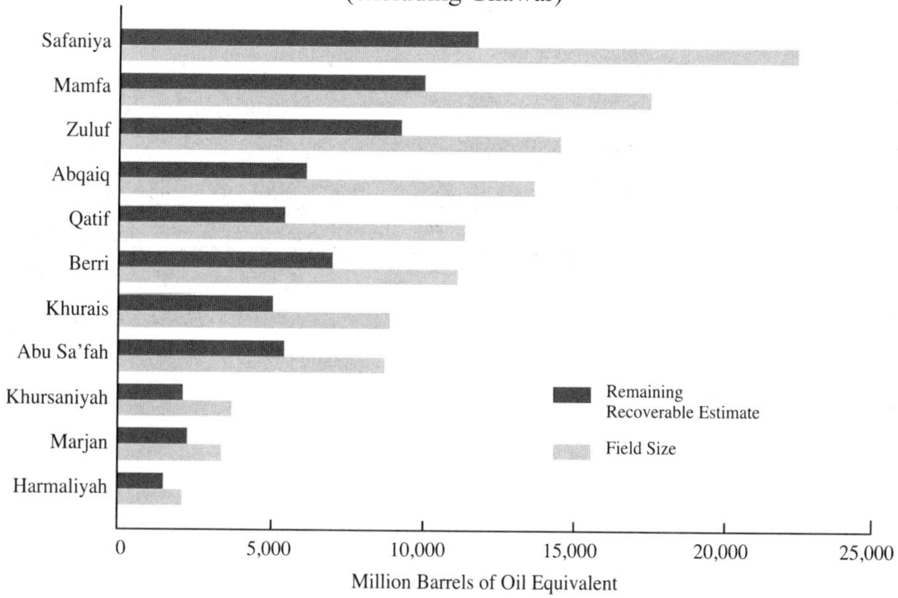

Note: Fields indicated in this figure are oil fields including crude, NGLs and condensates.
Source: Cambridge Energy Research Associates (CERA).

CERA does not share analyst Mathew Simmons' concerns about Saudi Arabia's ability to increase production capacity. With these levels of reserves and resources, it is CERA's view that Saudi Arabia will be able to maintain and increase its capacity in the short to medium term (to 2020) if reservoirs are properly managed, government policy is stable and there are no unforeseen geopolitical upheavals. According to CERA analysts Robert W. Esser and Peter M. Jackson, Saudi Arabia's most likely production profile shows a gradual capacity increase to about 12.8 mbpd by 2015 (see Figure 4.3).

According to media statements made by Saudi Aramco, both a 12 mbpd and a 15 mbpd capacity can be sustained for 40 to 50 years. It should be noted that the Kingdom's oil policy is driven by economical as well as by political imperatives.

Figure 4.3
Saudi Arabia's Liquids Capacity

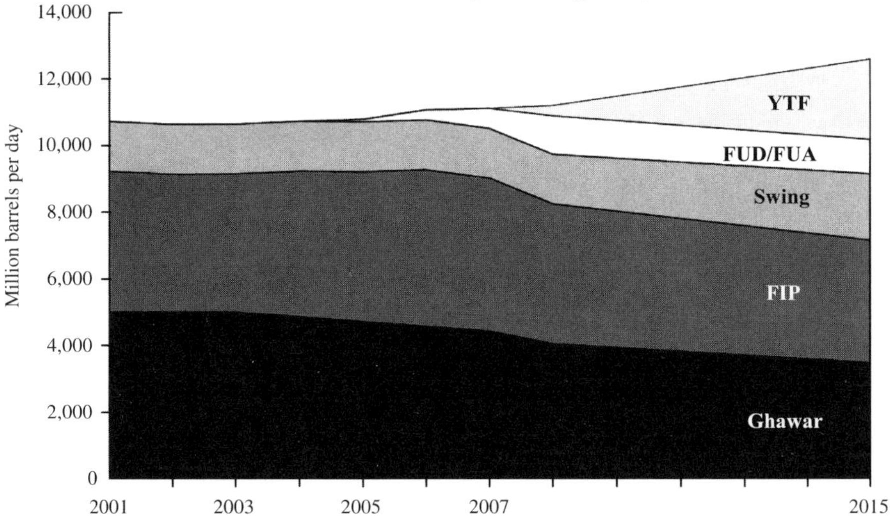

YTF: yet to find   FUA: field under appraisal
FUD: field under development   FIP: field in production

Source: Cambridge Energy Research Associates (CERA).

Maximization of Oil Revenues

- *Oil provides the bulk of Saudi revenues:* This proportion varies between 73 and 85 percent of the country's revenues depending on the year (see Figure 4. 4).
- *Oil wealth is part of the implicit social compact:* As long as the state is seen as the ultimate protector, which can ensure that the basic needs of its population are satisfied, the ruling family can maintain a position of total political control. However, the day the state is forced to levy direct taxes, the issue of the government's accountability toward the population will be raised.
- *Dramatic demographic growth is raising the stakes:* Demographic growth is a key factor that will determine the ability of the state to maintain a steady, and if possible growing, flow of revenues. According to CERA estimates, the Saudi population has doubled between 1980 and 2000 and despite a slowing down of the demographic annual growth

rate from 4.2 percent in the late nineties to 3.2 percent, it is expected to double again to 33 million in 2020 (see Figure 4.5).

Figure 4.4
Oil's Share in Saudi Arabia's Revenues

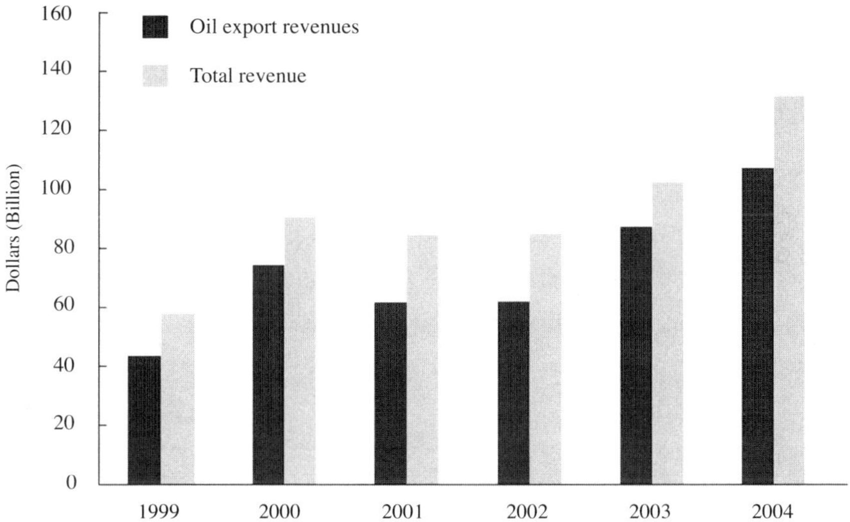

Source: Cambridge Energy Research Associates (CERA).

Figure 4.5
Saudi Arabia: Population Growth of Nationals

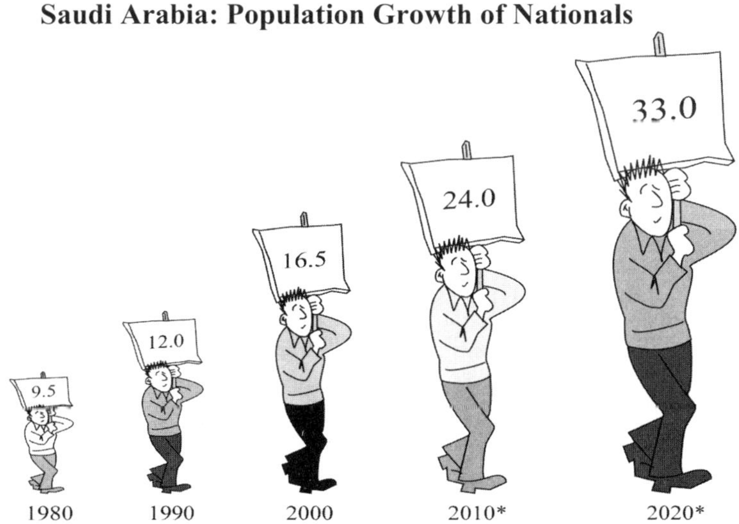

Source: Cambridge Energy Research Associates (CERA).

According to the census conducted in late 2004, population growth in Saudi Arabia has slowed down more than expected. Population growth has averaged 2.49 percent per year between 1992 and 2004. Hence, the Saudi population is now expected to reach 30 million by 2030.

Political Clout

There is a very close correlation between Saudi dominance in the oil sector and its role on the world scene. There can be no doubt therefore about the importance of dispersing any doubts regarding the political stability of the kingdom.

- Saudi Arabia's regional weight, but more so, its importance for the United States, are both factors that are directly linked to its position as the top oil producer and exporter.
- The government of Saudi Arabia has always paid attention to maintaining its US market share. Saudi Arabia has been and hopes to remain one of the four main suppliers to the United States. In August 2004, the Kingdom even regained, at least on a monthly basis, the top position (see Figure 4.6).
- While Saudi Arabia's reserves and exports capacity are much higher than those of Russia, the spectacular re-emergence of Russia on the oil scene and the intended cooperation between Washington and Moscow in the energy sector was viewed for a time as presenting a potential for serious competition (see Figure 4.7). However, recent developments in Russia seem to have significantly slowed down its potential oil capacity growth.
- Saudi Arabia's role has been significant in OPEC history, for example, in the November 1997 Jakarta conference that decided to raise the OPEC production ceiling and consequently sent prices spiralling down. This particular decision stemmed from the competition for market share between Saudi Arabia and Venezuela. Conversely, by achieving a consensus between Saudi Arabia, Venezuela and Mexico a few months later, OPEC was able to reverse the price trend.

Figure 4.6
Top Five US Oil Suppliers

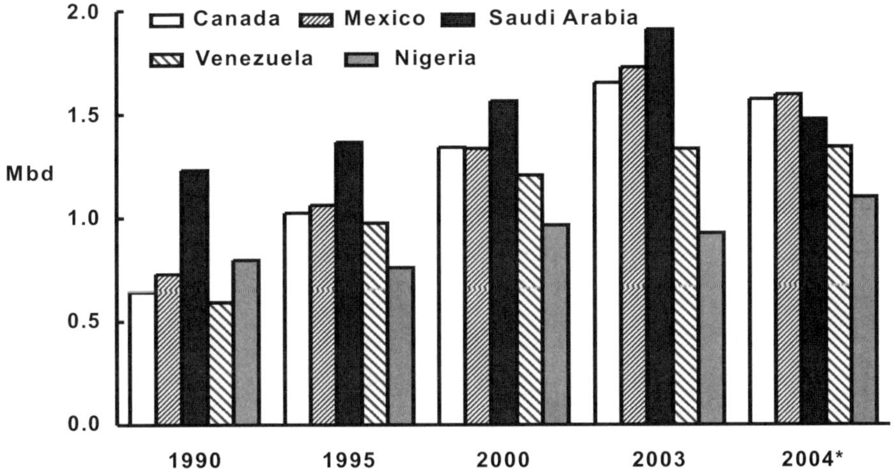

Source: Cambridge Energy Research Associates (CERA).

* Estimates

Figure 4.7
A Slowdown in Russia's Production Capacity

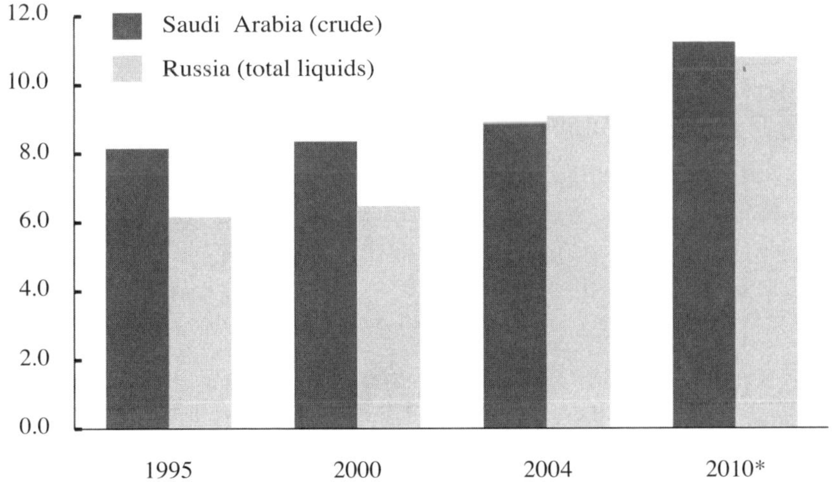

Source: Cambridge Energy Research Associates (CERA).

* Estimates

OPEC Leadership Remains a Key Driver of Saudi Policy

OPEC's importance for Saudi Arabia is far more than just market share. Saudi Arabia derives part of its political influence from its role in OPEC's price-setting policy. Ever since the 1973 oil crisis and the oil embargo (which was a decision by OAPEC and not OPEC), it has been Saudi Arabia's strategy to establish itself as a reliable source of supply for the benefit of both the producers and the consumers.

Recent events have shown that calming down sky-rocketing oil prices might prove to be as delicate an exercise as trying to prop up depressed oil markets. Only countries with a sizable spare capacity can even try to influence price trends. The current problems stem from the fact that, in the face of many potential supply disruptions, OPEC's spare capacity has been reduced to about 1.2 mbpd – less than 1.5 percent of world consumption – and that this spare capacity is almost entirely in the hands of Saudi Arabia.

Securing the Role of Oil in the Energy Mix

As a country with reserves expected to last 78 years at current production levels, Saudi Arabia's concern is not only to secure its market share today against non-OPEC supply, but also to preserve the important role of oil in the global energy mix.

- It means ensuring that very high oil prices do not encourage investments in non-OPEC countries and do not choke off demand.
- It means taking a pro-active attitude as far as environment regulations are concerned.
- It also means keeping a close eye on alternative energy sources in the transportation sector, especially fuel cells.

In order to achieve these objectives, Saudi Arabia is undertaking the following measures and initiatives:

- Saudi Arabia is relying on a national champion, Saudi Aramco, which enjoys a monopoly on upstream oil. While policy-making and regulations are set by the Ministry of Petroleum and Mineral Resources, the national company is an independent organization and is managed along the lines of any other international company.

- Saudi Aramco strives to develop a state-of-the art industry. Built on the inherited structure of the former Aramco, the national oil company focuses on efficiency and the in-house development/or the acquisition of the latest technologies.
- As Saudi Arabia's Minister of Petroleum and Mineral Resources Ali Naimi observed at the OPEC International Seminar in Vienna, 2004: "Saudi Aramco's challenge is how to mobilize its own and cross-industry resources to promote the use of oil and gas. Carbon sequestration and capture technologies, as well as technologies that ensure a role of these fuels in hydrogen generation, are emphasized in the R&D programs of the company and in its collaborative research with the industry, universities and research institutions."[2]
- Saudi Arabia is securing outlets in its major markets by investing in their downstream sectors. Taking into account that the bulk of the demand growth is expected to keep coming from Asia, Saudi Arabia has shifted its downstream priorities from the United States[3] to South Korea,[4] Japan and China.
- To preserve its unique position in the oil market, the Kingdom has to maintain its usual 1.5–2 million bpd spare capacity cushion.

The main challenges confronting Saudi oil policymakers are as follows:

- Choosing the right time for the development of new fields. Bringing additional production too quickly on stream can prove a costly exercise. Conversely, failure to have an adequate cushion of spare capacity risks triggering a major oil crisis and being eventually detrimental to long-term demand for Saudi oil. As the CERA report by Jackson and Esser observes, bringing new capacity on stream takes between 18 to 24 months, but assessing the medium term supply/demand balance is a tricky exercise.
- Fine tuning OPEC production management in such a way that oil prices do not, in the long run, affect demand and/or encourage non-OPEC supply.

- Making sure that Saudi Aramco's social responsibilities do not override sound investment decisions. While Saudi officials insist that Saudi Aramco belongs to a different class of national oil companies (NOCs), inasmuch as there is a clear separation between policy-making, regulations and operations, the danger remains that Saudi Aramco's role in diversifying the economy and developing human resources might lead the company to take decisions that may be seen as detrimental to its own interests.

Saudi Arabia's Gas Industry

At the outset, it must be noted that Saudi Arabia does not enjoy the same supremacy in the gas sector as it does in the oil sector (see Figure 4.8).

Figure 4.8

Middle East and Former USSR: Ultimate Recoverable Reserves

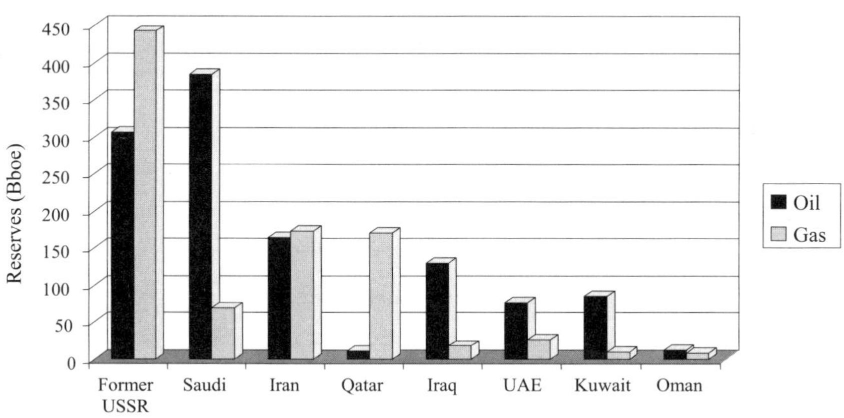

Note: Proven plus Probable (P+P) Estimates
Source: IHS Energy

Saudi Arabia ranks fourth in terms of gas reserves and holds the eleventh position among the gas producers in the world. This has led the Kingdom to differentiate its gas approach from its oil policy. The only common point is that both sectors are not labor-intensive industries (see Figure 4.9).

Figure 4.9
The Saudi Challenge

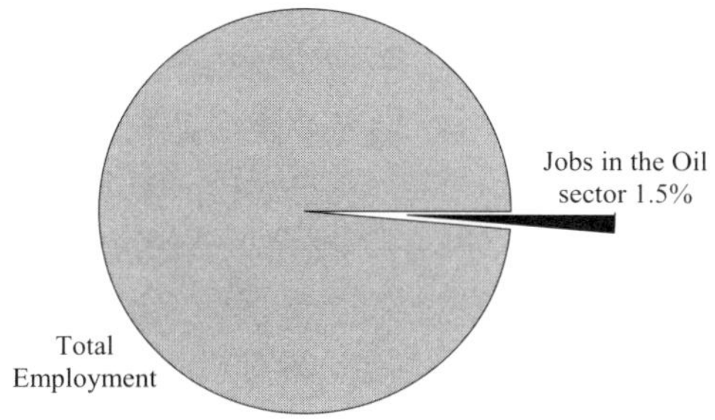

Source: Cambridge Energy Research Associates (CERA).

It was a fallacy to present the now defunct "Gas Initiative" as a direct source for numerous job opportunities. However, gas development remains a key factor to further economic development in Saudi Arabia.

The recent focus on gas is driven by:

- A surge in domestic demand that is poised to grow by 5 percent per year from the current 7 billion cubic feet per day to 14 billion cubic feet per day in 2025 (see Figure 4.10). Saudi Arabia consumes more gas per capita than the highest income countries in the Organization for Economic Cooperation and Development (OECD).
- There is a growing demand for basic needs, water and power. According to CERA sources, the level of water consumption is set to almost double within 20 years, from 2.3 billion cubic meters per year in 2003 to 4 billion cubic meters per year in 2020, whereas power demand will be multiplied by 2.4 over the same period (see Figure 4.11).
- It is clearly important to develop an industrial base fuelled by gas, since gas-based utilities and industries account for 15 percent of GDP in Saudi Arabia. There has recently been a surge in Saudi petrochemicals projects, as shown by the Memorandum of Understanding signed on May 9, 2004, between Saudi Aramco and Sumitomo for the development of a $4.3 billion refinery and petrochemical complex in Rabigh, and also by the $3 billion Project Management and Development (PMD) venture for a Jubail-based feedstock cracker, announced early in 2004.

Figure 4.10
Saudi Arabia's Domestic Gas Demand

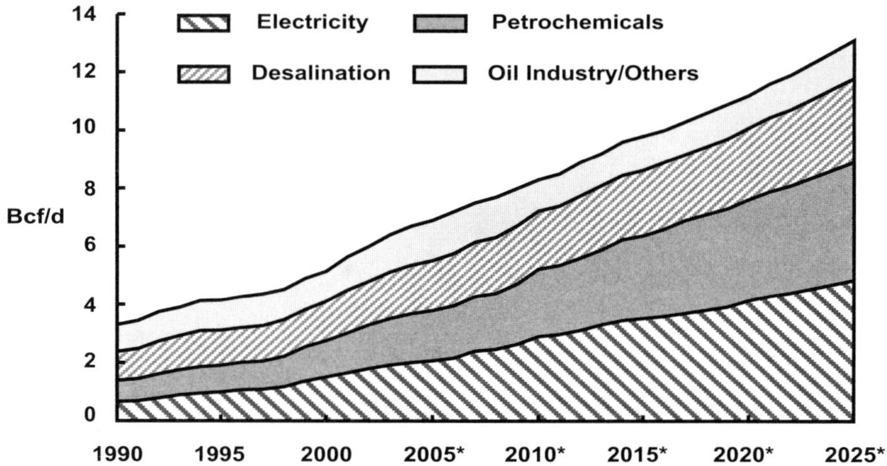

Note: * Indicates forecasts.
Source: Cambridge Energy Research Associates (CERA).

Figure 4.11
Saudi Arabia: Growing Demand for Water and Electricity

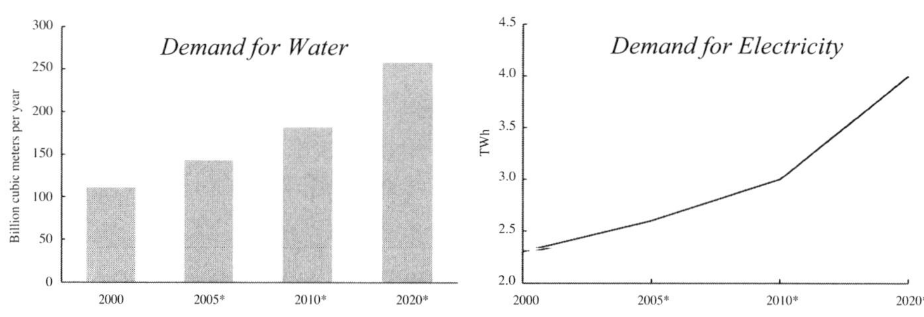

Source: Cambridge Energy Research Associates (CERA).

The implications are that Saudi Arabia has to carry out the following:

- Expand current proven reserves of 235 trillion feet of gas reserves through an aggressive exploration program.
- Increase the share of non-associated gas in order to avoid boosting oil production and meet, OPEC quota constraints, as done in the past.

Due to its associated gas requirements, Saudi oil production cannot go below 7 million bpd (see Figure 4.12).

- Take confidence-building measures in order to convince the petrochemical industry that its feedstock needs will be met by adequate gas supply.
- Address the concerns of the petrochemical industry about liquefied petroleum gas (LPG) prices for the local industry as compared to the export prices. Some trading partners have objected to this situation as constituting "dual pricing."
- Allow foreign investments for gas exploration activities so as to free up public funding for growing social requirements.

Figure 4.12
Saudi Arabia's Gas Reserves (235 Tcf)

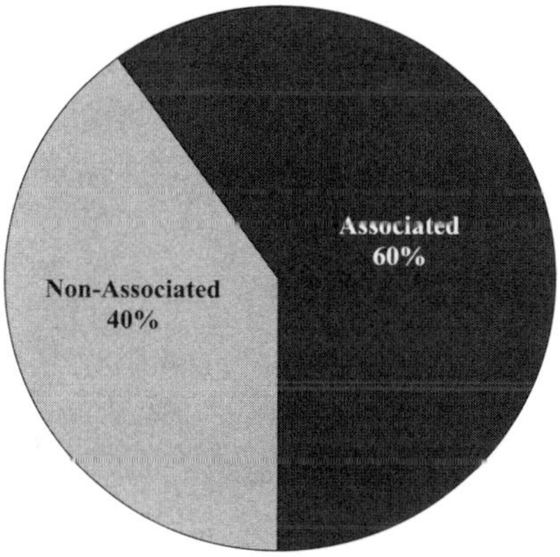

Source: Cambridge Energy Research Associates (CERA).

This different set of issues has led to a substantially new playing field with:

- The completion of the Master Gas System (MGS), which comprises processing plants and pipelines that deliver clean gas to utilities and petrochemical plants. The MGS handles over 6 billion cubic feet of gas per day through the Al Barri, Shadqam, Al Uthmaniyah, Hawiyah and Haradh facilities (see Figure 4.13).

- There has been a long-awaited partial opening of the gas upstream sector, but in partnership with Saudi Aramco, which holds between 20 and 30 percent of all the companies set up to handle the ventures.
- A set of Gas Investment Tax regulations was promulgated in January 2004.
- A new set of players has been introduced into the field, with European companies (Total and Shell in the southern part of the Rub Al Khali (Empty Quarter); ENI and Repsol in the northern part) but also newcomers in the region (Lukoil, CNPC and Sinopec) being involved in the northern part of the Empty Quarter (see Figure 4.14).
- There has been a withdrawal of the midstream and downstream projects that formed part of the Gas Initiative.

Figure 4.13
Saudi Master Gas System

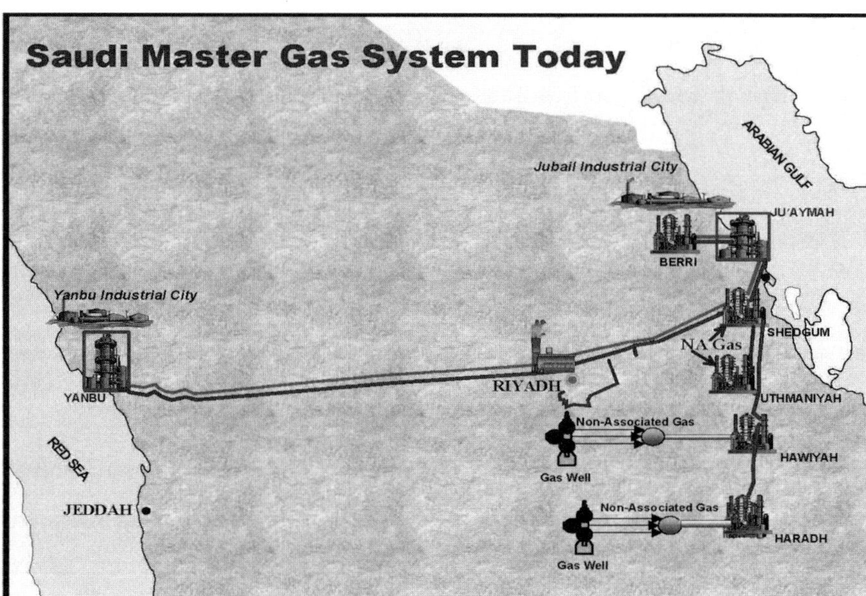

Source: IHS Energy

Figure 4.14
Saudi Gas Upstream Contracts with Foreign Companies

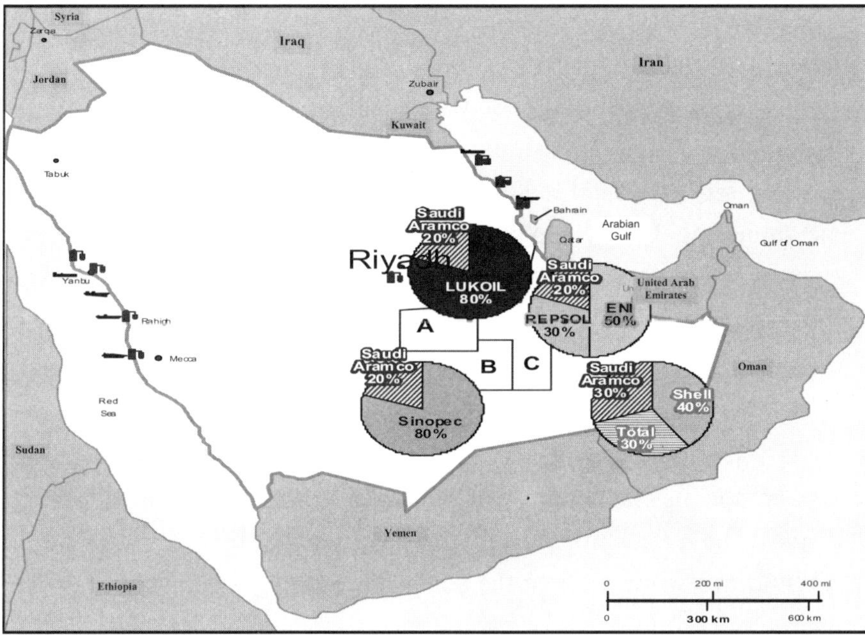

Source: Cambridge Energy Research Associates (CERA).

Saudi Arabia's priority is to monetize gas by expanding its domestic market. However, at one point, it will have to face a challenge and decide whether to explore gas export opportunities by LNG or choose the gas-to-liquids (GTL) path or perhaps both. The decision will obviously depend on three different sets of factors:

- *Size of reserves:* One consideration is the additional amount of gas that will be discovered in Saudi Arabia in the near future versus the size of the domestic market. In this regard, some question the wisdom of restricting exploration by foreign companies to the Empty Quarter, while many IOCs have pleaded for access to the Saudi Aramco Reserved Area (SARA) and wonder how empty the Empty Quarter will prove to be. The exploration results will enable the authorities to decide whether to pursue the development scheme on both fronts – domestic market and exports – or not.

- *Economic Factors:* Assuming that the only export route would be via LNG, is it more interesting economically – in terms of revenues – to export Saudi gas or to sell liquids?
- *Commercial Factors:* In addition to the growing domestic demand, Saudi Arabia will have to take into account strong competition from neighbouring Qatar, Egypt and Iran: first, in the race for foreign capital and second, for marketing its gas, should it decide to target foreign markets.

Conclusion

To sum up, the biggest challenges that Saudi Arabia must deal with are to:

- Choose the right timing for its oil capacity expansion
- Calm down worries about the adequacy of the gas supply to domestic markets
- Avoid being side-tracked by competition for gas exports in the event of significant new gas discoveries.

5

The Norwegian Experience of Economic Diversification in Relation to the Petroleum Industry

Øystein Noreng[*]

From the early 1970s to the mid-1990s, the Norwegian government imposed on foreign oil companies the requirement to use domestic goods and services, to invest in joint industrial ventures and to transfer knowledge. The industrial venture policy was unsuccessful, but the procurement policy quickly raised the local content to high levels and the knowledge policy contributed to technological change that considerably reduced costs and established Norway as one of the leaders in petroleum technology. The programs were sponsored by the government within a regime of high marginal tax, with the Treasury bearing the major part of the cost. In hindsight, the issue is whether this infant industry protection was necessary or whether market forces would have been able to produce comparable results as quickly.

Prior to 1970, Norway had no indigenous oil or oil services industry; but there was consensus that it had to develop one. As market forces alone were seen as insufficient, government intervention was considered indispensable. At an early stage, licensing included state participation, a national oil company, Statoil was established, and legislation was

* The author is indebted to his colleague Professor Svein Andersen of the BI Norwegian School of Management and his former assistant Jostein Dahl Karlsen, currently with the Norwegian Ministry of Petroleum and Energy (who has authored a thesis on Norway's experience of industrial cooperation with the international oil industry) as well as to INTSOK for valuable advice.

introduced requiring oil companies to use Norwegian goods and services whenever competitive, followed by provisions to transfer competence and cooperate in the development of new technology. Foreign oil companies were enticed to engage in industrial ventures unrelated to petroleum and in joint ventures with Norwegian research institutions to develop petroleum relevant technology. In 1994, the trade agreement with the European Union discontinued these policies. By then, the Norwegian petroleum cluster no longer needed infant industry protection.

In the use of the financial surplus, Norway has established a Petroleum Fund whose objective is to shelter the domestic economy from overheating and protect it from the vagaries of oil prices, as well as to build up financial assets as a means of income and risk diversification. There are no safeguards against the opportunistic use of the Fund, as the Parliament is constitutionally sovereign.

The Purpose

The purpose of this chapter is to review and discuss the Norwegian experience of economic diversification in relation to the petroleum industry. After discussing the background and rationale for economic diversification, it reviews the different policies of economic diversification chosen: procurement, transfer of knowledge and petroleum revenue investment. Finally, it discusses the significance of the Norwegian experience for other oil exporting countries, especially the Arab Gulf states.

The thrust is descriptive rather than analytical, aiming at explaining the different phases and aspects of the relevant policies, rather than analyzing their costs and benefits, a task that would exceed the scope of the present chapter. Nevertheless, some basic issues and questions cannot be avoided, such as the impact of government intervention versus reliance on market forces, the value of infant industry protection, the significance of the excessive dependence on unstable resource rents stifling growth in other sectors, and the eventual impact of institutions and policies.

From the early 1970s to the mid 1990s, Norwegian petroleum policy represents a fairly systematic case of infant industry protection. Traditionally, infant industries have been protected by tariffs.[1] For a generation, Norway's government regulations may have been more effective as tools of microeconomic intervention than tariffs.

The argument in favor of protecting an infant industry is usually the need for temporary sheltering of a new activity that in the short run would be unable to survive competition with established foreign suppliers, but which has the potential to become competitive. Consequently, the reasoning is dynamic, the bottom line being that the need for protection will diminish over time. From this perspective, infant industry protection is a transient arrangement whose social costs in the short run are supposedly offset by social gains in the longer run. This would require the increase in future national income to exceed what it would have been if the infant industry had not been the subject of protection. The traditional argument is that infant industry protection is warranted for small new firms that have little chance of competing with established firms as they have been in business longer and over time have been able to improve their efficiency. Historically, they are likely to have developed competitive advantages, exploit economies of scale and have gained superior knowledge of markets and technology.

The argument is mostly used in conjunction with establishing new industries in developing countries, to justify sheltering against established competitors in industrial countries. The justification is also made in conjunction with alleged imperfect competition. This was the case with the Norwegian oil industry around 1970. The choice in Norway was not a tariff, but provisions on local content that within a short time established a large captive market for petroleum-related goods and services. The very success of the infant industry protection policy may inadvertently have enhanced the risk exposure of the Norwegian economy, indirectly strengthening the resource curse.

The established hypothesis, put forth by Jeffrey D. Sachs and Andrew M. Warner, is that the resource curse with a high risk exposure is practically unavoidable for countries whose governments have access to economic rent, as their expenditure, after adjusting for other relevant factors, causes significantly higher prices in resource-dominated economies, compromising the competitiveness of the non-resource-based parts of the economy.[2] The "Dutch disease" is the classic example.[3] It means that the discovery and exploitation of natural resources de-industrializes a nation's economy, as costs rise, the value of the country's currency rises, making manufactured goods less competitive, imports increase and non-resource exports decrease. The outcome is high risk exposure, instability and precarious public finances and employment. It is relevant to the oil exporters of the Middle East and Venezuela.[4] A contrary hypothesis, put forth by Mehlum, Moene and Torvik, is that countries with large natural resource endowments constitute both growth losers and growth winners, the main reason being differences in the quality of institutions.[5] Apparently, this should apply to Norway, together with countries such as Australia and Canada.

Experience varies. The coexistence of petroleum revenues based on economic rent and industrial income is difficult. In some cases, such as Mexico, with idle human resources, petroleum revenues can provide capital for industrial investment, strengthening employment.[6] In other cases, such as the Netherlands, with less available human resources, using rentier revenues can have a crowding-out effect, causing loss of industrial jobs in sectors exposed to international competition.[7] In Norway, a tight labor market was early on identified as the major constraint on spending petroleum revenues.[8]

The "Dutch disease" and the resource curse are not the themes of the present chapter, but they are relevant to Norway's attempts to diversify the economy away from excessive dependence on petroleum revenues. Indeed, it was the resource endowment that both spurred and facilitated the industrial policies concerning petroleum. The local content, the knowledge transfer and the industrial employment policies imposed by the Norwegian government on foreign oil companies can be seen as ways to capture economic rent in addition to the tax-paid cost, even if the government through the high petroleum taxation in the last instance bore the major part of the expenses. In hindsight, it is an open question as to what extent the apparent successes of Norway's petroleum policy have been offset by less apparent social costs. Government intervention in industry is seldom without flaws and often leads to inefficiencies and distortions, at times even to outcomes contrary to intentions.

The value derived by society in supporting and protecting infant industries by investing in human capital and limiting foreign competition has been an issue among economists for generations, and opinion is still alternating between reliance on market forces and government intervention. In the case of Norway, the question is whether the use of local goods and services and the training of nationals would have evolved in any case on the basis of competitive locations and skills, but possibly at a lower cost through the market than through state intervention. The question is also whether market forces would have produced comparable results as quickly.

A final question concerns the lessons learned from the Norwegian experience of economic diversification that could eventually be applied to the member countries of the Gulf Cooperation Council, GCC. The issue of substantially reducing basic dependence on oil revenues is pertinent also in the Gulf. The challenge therefore is to establish real economic linkages between the oil and gas extraction and other sectors. The business clusters such as Houston, Calgary, Aberdeen and Stavanger in Norway indicate the potential for spin-off secondary and tertiary economic activities from oil and gas extraction, and that there are parts of the petroleum value chain above the upstream activities, through the supplies of goods and services, serving markets outside the local and national arenas. The issue is how the Norwegian model of economic diversification eventually could be applied to the GCC countries in shifting their current focus from merely extracting gas and oil resources to developing a broader economic base.

The Background

Around 1960, discoveries of natural gas in the Netherlands alerted the international petroleum industry to the North Sea's petroleum potential. In 1962, the first oil companies addressed the Norwegian government requesting permission to explore for petroleum on the continental shelf. At the outset, being approached by the international oil industry in the early 1960s, the Norwegian government realized that it lacked the competence and insight to bargain.[9] Instead of striking a quick deal, it preferred to study the matter.[10] The first insight was that oil should be handled with care because of the value involved and because the international oil industry at the time to a large extent was not perfectly competitive, but rather an oligopoly, vertically integrated with less than fully transparent transfer pricing, and with close links to the major oil importing governments.[11] The apprehension was that the major international oil companies, cooperating in a *de facto* cartel, were enjoying excess profits at the expense of consumers and oil exporting countries alike.[12] Against this backdrop, the exercise of state power over oil and gas in Norway embodied a competitive strategy.[13] Acting in this way, the Norwegian government assumed the role of a mercantilist state, tying oil to other interests in a context of imperfect markets, business conflict and rent seeking, following an old tradition in the oil business.[14]

Norwegian consensus at the time was that markets are efficient for some purposes but not for others.[15] In the Norwegian context of the 1960s and the 1970s, a liberal market-based approach to the petroleum industry would hardly have been feasible. There is continuity from hydroelectricity policies established early in the 20th century, aiming at keeping national control, to the hydrocarbon policies decades later, also aiming at national control and keeping a strong bargaining position for the government landowner with foreign investors. The division of the economic rent, the surplus above the normal, risk-adjusted return on capital, was not the only issue, but also the integration of the petroleum industry into Norwegian society, meaning its use of local personnel, goods and services. This was a prerequisite for social and political acceptance, and eventual additional costs incurred were mostly considered irrelevant or a necessary price. A few lone voices advocated auctioning off the oil prospects, leaving their development to the market, essentially foreign companies, but found little agreement among politicians. The trade unions preferred bargaining with Norwegian employers; the fishing interests were at first skeptical to the petroleum activities and had a preference for Norwegian companies that were assumed to be more respectful of environmental concerns; the business community likewise had a preference for dealing with Norwegian oil companies and many realized that the petroleum industry value chain commences before the traditional upstream, with goods and services.

At the time, the case for strong government intervention in the emerging petroleum industry was also made by reference to market failures. The point for Norway at the outset was to hedge a bargaining position with the international oil industry that was seen as less than fully competitive, dominated by an oligopoly of large companies, often with close links to the governments of oil-importing countries.[16] There was considerable anxiety that a petroleum industry dominated by foreign companies would have a preference for their established suppliers of goods and services, to the detriment of Norwegian newcomers.

The international context provided additional arguments: Norway's petroleum policies were developed at a time when the member countries of the Organization of Petroleum Exporting Countries (OPEC) asserted their power by quadrupling world market oil prices and subsequently nationalizing most upstream assets. Consequently, as Norwegian oil policies developed in the 1970s, there was a dramatic strengthening of the bargaining power of oil exporting countries over the international oil industry. This evidently benefited Norway, with a favorable location, interesting resource prospects and political stability. Norway promptly declared common interests with OPEC in matters of oil markets and prices.[17] The drawbacks were a difficult environment, deep waters, inclement weather conditions, technical obstacles and high costs. Since then, technology and organization have improved and costs have declined, and the Norwegian petroleum industry has moved into deeper and more northerly waters. During the 1990s, the cost decline in the Norwegian offshore petroleum industry was impressive, apparently more substantial than in UK waters or in the Gulf of Mexico. It is an open question as to the extent to which this was due to the massive research effort or to the end of protectionism.

The Norwegian story is about microeconomic change through a persistent cost pressure and public policy, macroeconomic challenges from large and unstable petroleum export revenues, and a transformation of the political economy where the state has emerged as the major accumulator of capital, and consequently, the leading investor. The story is also about a state that is active in industrial policy, exercising authority and control on the basis of ownership of natural resources, the limited competence of markets in handling strategic and structural issues, and finally, an apparently successful combination of the state and the market. As a result, the interaction between the government and the international oil industry has been more balanced and less conflict-ridden than in many developing countries.[18]

The Rationale

From the outset, knowledge and insight were major concerns among Norwegian policy makers. The apprehension was that successful bargaining with the international oil industry would require insight into the details of the operations, from exploration through field development, extraction and refining, including geology, technology and finance. The view was that the best way to gain insight would be active participation in all aspects of petroleum activities. Against this backdrop, provisions for state participation were introduced in the 1969 licensing round. As the bargaining position improved in the early 1970s, foreign oil companies were relegated to a secondary position. The costs of Norwegian dominance were hardly a matter for discussion.

The principle of state participation and the need for a strong national oil company were hardly discussed, but there was some disagreement on how to organize the national company. In any case, Statoil, the national oil company, was established by a unanimous parliament in 1972. The intention was to make the company a compatible bargaining partner for the large international oil companies. The main task was to develop the resources on the Norwegian continental shelf. The organizational model chosen for developing infrastructure was hierarchical, with Statoil at the top of the command chain, integrating decisions and coordinating the various projects. This position provided Statoil with organizational and financial resources, economic leverage and political influence. To some extent, with a government owner not demanding dividends but fulfilling different objectives, the company had incentives to seek power and influence rather than efficiency and profits. The position as a dominant actor gave incentives to rent-seeking behavior.[19] Several cost overruns and subsequent changes of senior management indicate that the government control of a large, wholly state-owned oil company did not work satisfactorily, which is a problem not confined to Norway or Statoil.[20] One objective of the partial privatization of the company was to make Statoil subject to the scrutiny of financial markets as a way to control costs and enhance transparency.

A consistently declared Norwegian policy has been to diversify in order to reduce dependence on petroleum revenues and the ensuing risk exposure to oil and natural gas market instability. Economic diversification away from petroleum has not been an unmitigated success. The background of the policy is simply that when petroleum activities commenced in the 1960s, Norway already was a developed industrial country with a strong tradition in shipping and hydro-electricity, as well as a diversified manufacturing base, including capital and energy intensive metal and wood processing industries, shipbuilding and mechanical industry. In the

1960s, Norway enjoyed comparatively strong economic growth, full employment and a current account surplus and therefore had no urgent need to rush ahead with petroleum development.[21] This comfortable economic situation provided the basis for a strong bargaining position with the international industry, in contrast to most developing countries. Evidently, Norway's bargaining position improved as oil was found and as prices shot up in the 1970s.

Higher oil prices also created concerns.[22] Before the first oil price increase, in the spring of 1973, the assumption in Norway's Finance Ministry was that developing the industry aiming at an annual output of about 120 million tons of oil would raise the economic growth rate from 4.5 to 5.5 percent through demand for goods and services and public revenues. Concern was to hedge the bargaining position through insight and competence considered necessary to control the industry. This objective led the government to impose other obligations than just taxation on the oil companies.

The rationale was briefly that the petroleum industry operates on public land, and extracts resources-in-the-ground that are in public ownership.[23] In Norway, as in most other countries, resources in the ground are state property.[24] Consequently, lifting oil and gas means depleting a public capital base, for which compensation should be found in the building up of other assets to secure a continuity of income. The challenge for the government-landowner is at the minimum to regulate activities, eventually to act as an entrepreneur in handling natural resource wealth.[25] Norway, like the UK, practices discretionary licensing, awarding licenses upon bargaining with oil companies over multiple objectives, not only to maximize revenues, which stresses the mercantilist character of the policies chosen.

Oil and natural gas in the ground have an intrinsic value, although variable, because company costs are in exploration, development and lifting, but not in creating the raw material that has a market value, providing the basis for an above-normal return on capital, which essentially is the resource rent.[26] By restricting access, the government-landowner also raises barriers to entry and limits competition, ensuring that investors not only catch a resource rent, but also an oligopoly rent from imperfectly competitive markets. The division of the rent is a bargaining matter between the government-landowner and the companies and in Norway, the total government "take" (the sum of revenues accruing to the government) is up to 78 percent. The outcome is typically special petroleum taxation, often in combination with production sharing agreements (PSA), to capture most of the rent, as well as other provisions that the government-landowner is sovereign to request.[27] Petroleum exploration and taxation affect other industries through cost pressures and conflicts over uses of acreage, especially with the fishing industry. The North Sea record is essentially positive on this account. Policy decisions in other areas affect the petroleum industry (such as fishing, health, labor relations, industrial safety and education).

The government-landowner can demand more than money, depending on the situation and bargaining power. The rational solution is to develop a local supply of some competitive goods and services required by the petroleum industry, not to demand an indiscriminate use of local suppliers. The risk is that requirements to the petroleum industry become a balancing factor in regional, industrial and employment policies, so that the petroleum industry becomes a substitute for the capital market, creating jobs that are not competitive and not sustainable.[28] The risk is also that the petroleum industry will cause a local cost pressure, crowding out other economic activities.

Transfer of knowledge can be in both parties' interest. The government-landowner has a legitimate interest in enhancing its bargaining position through the development of competence and the knowledge base. The local community has a legitimate interest in competence growth to compensate for resource depletion. The oil industry will need local personnel and skills—a local constituency. The scope and methods of the knowledge transfer are negotiable.

The key is focus and selectivity. The government-landowner should not and cannot demand everything. One solution is to focus on revenues, ignoring other possible benefits. The alternative is to take some other benefits, but marginally at the expense of revenues. The issue is therefore who realizes and distributes what benefits the most effectively—the government or the oil companies. The time-horizon for the petroleum activities is decisive for the balancing of monetary and non-monetary benefits. A comprehensive perspective is required for the government to fulfill its role as a caretaker of overall economic and social concerns in relation to the petroleum industry. This has been a policy objective in Norway since the 1960s, but practice has not always been equally successful.

The opportunities include enhanced revenues to the Treasury, permitting an increase of expenditure without corresponding taxation, and permitting the growth of both consumption and investment. Oil revenues bring resources to develop infrastructure and competence. New business development and employment are strengthened through oil industry procurement of goods and services. Training and transfer of knowledge can take place through joint ventures of oil companies and local businesses. Establishing business clusters for the oil industry is also important. Neglecting the development of business clusters through bureaucracy and inertia would mean ignoring the need for a dynamic private sector. On this account, Norway has succeeded, but not without problems.

The Norwegian success in achieving high local content is largely due to government policies, which encouraged partnerships between foreign and domestic companies, and made research programs mandatory. This ensured that technology developed in Norway would be among the best of its kind. Since 1970, successive governments have regarded it as essential to promote competition in the oil industry, while at the same time actively promoting the business opportunities for Norwegian industry.[29]

The risks and the pitfalls are also numerous. They include not collecting potential taxes through mismanagement of the bargaining situation.

Squandering the oil revenues on private consumption, social and regional programs, at the expense of investment in more durable assets, such as infrastructure and training, is another risk. Neglecting the market for local goods and services in the oil companies, or practicing undue protectionism, are well-known problems in oil producing countries. Neglecting knowledge development means ignoring the importance of human capital. On this account, Norwegian policy has perhaps been the most successful.

The windfall profits from oil and natural gas development also present risks. Because oil prices usually contain an economic rent, there is little, if any, relationship with real supply costs, and consequently, they are unstable.[30] A common scourge of oil exporting countries is that a sudden influx of oil revenues, in the wake of an oil price increase, at first leads to a combination of local inflation and currency appreciation, compromising the competitiveness of the non-petroleum sectors. Subsequently, as oil revenues decline or are spent, an economic squeeze occurs, with the need to cut consumption and investment and to depreciate the currency in order to enhance non-petroleum competitiveness. In this way, many oil exporting countries experience cycles of boom and bust, victims of the vagaries of the international oil market.[31] The alternative is to use petroleum revenues with caution and preferably in an indirect way. Since the early 1970s, Norway has had its share of booms and busts, and only since the early 1990s have macroeconomic policy and the use of petroleum revenues been more sober. Norway currently has a Petroleum Fund, Europe's largest investor, with assets of about $150 billion by mid 2004, invested in bonds and the stock market.

Norwegian oil and gas production has increased significantly over the past three decades, and Norway ranks today as the world's third largest exporter of crude after Saudi Arabia and Russia. The petroleum industry in 2004 accounted for about 20 percent of Norway's gross domestic product (GDP), about half of exports, about a third of government revenues, but only 1.2 percent of employment. Petroleum production and pipeline transport create substantial revenues for licensees and the government. Through its demand for goods and services, this sector also generates substantial activity. The petroleum business has thereby been a prime mover in developing such sectors as mechanical engineering and associated services.

High petroleum taxation – up to 78 percent of net income – means that the government as the tax collector assumes most of the cost of the various schemes imposed on the oil companies, whether procuring local goods and services, training locals or participating in research.

Procurement Policy

An important policy objective for the Norwegian government in relation to oil and gas resources has been, and still is, to lay the basis for developing a competent and viable petroleum-related industry. Acting in this way, the Norwegian government has assumed the role of a micro-economic manager, through an interventionist industrial policy.[32] It built on strong traditions in Norway.[33] In earlier times, the transfer of expertise from abroad and the build-up of domestic operations were important elements in Norwegian petroleum policy. The country now has a competitive petroleum industry, embracing a large number of Norwegian suppliers and operations covering most stages of the value chain from exploration via development to production and decommissioning.

In some fields, Norwegian suppliers to the oil and gas industry are among the world leaders. That applies particularly to seismic surveying, subsea installations and floating production systems. Activity in Norway's petroleum supplies industry has so far been related largely to new investment, maintenance and operational assignments on the Norwegian Continental Shelf—the NCS. The prospect of a lower level of capital spending in these waters means that the industry is increasingly focusing its attention on international market opportunities.

In the early 1960s, as mentioned, Norway had no indigenous oil industry. However, around 1970 there was consensus on the need to create one, with the help of the government. At the outset, a favorable economic situation caused a strong bargaining position with the oil industry. The use of local goods and services was explicitly ensured by law between 1972 and 1994. At times, the Norwegian share was above 90 percent of input. Norway has a large oil service industry that today appears oversized. Norway has established itself as a leader in offshore petroleum technology, especially in deep water, subsea and remote control operations. It has developed petroleum business clusters, but for some firms in labor-intensive activities such as primary mechanical construction, survival is at stake, due to high Norwegian labor costs and foreign competition.

In the 1960s, Norway had a mature shipbuilding industry with many yards along the coast, building all kinds of ships—from large tankers to specialized vessels. The industry also included a number of maritime equipment suppliers. Clients were Norwegian ship-owners, with one of the largest merchant fleets in the world. They recognized business opportunities in oil and the staff of well qualified sailors could be recruited to man rigs and supply vessels.

In 1967, the first small find was made, followed in 1969, by the first large find—Ekofisk. Also in 1969, the second licensing round took place. It introduced state participation on a carried interest basis, meaning that the state could enter a license group with 35 percent after a find had been declared commercial, and leaving the exploration risk to the companies. The state share was later expanded.

The Norwegian framework for the petroleum industry is based on the concessionary system, through which the companies get ownership of a share of a prospect. The licensing takes place through bargaining between the government and companies with the aim of agreeing on work programs. Licenses are always awarded to groups of companies. The purpose is to have different views and analyses, as well as risk-sharing through joint ventures, to provide the government-landowner with the maximum information.[34] There are multiple criteria for allocation, such as geological insight, technical competence, financial strength and performance record. The commitments are embodied in work programs and obligations to share insight. Experience from many developing countries was that international oil companies were often reluctant to share insight with the host governments.[35] From the very beginning, Norwegian legislation required oil companies to share all information with the government, and there was a bureaucracy able to enforce this rule.

In 1970, the government nominated a committee to review the organizational structure to take care of governmental interests and responsibilities. In the spring of 1971, the committee stated that in all oil producing countries the government tried to get better control through legislation and agreements with oil companies, and that these agreements included state participation in commercial discoveries. As a consequence of this more active policy, substantial organizations were established in many countries to take care of state interests and also state owned petroleum companies were established.

In 1972, the government was reorganized in relation to petroleum, with the establishment of the Norwegian Petroleum Directorate (NPD) as the administrative branch and the national oil company Statoil as the caretaker of the commercial interests. At the same time, a preference policy for Norwegian goods and services was introduced, accompanied by a policy of knowledge transfer and research cooperation.

To enforce procurement policy, in 1972, the Ministry of Industry established a Goods and Services Office as a watch-dog to control the oil companies' contracting and procurement activities. This was first applied with the third licensing round in 1973. Field development operators had to present to the Ministry a plan for all tenders above NOK 1,000,000 (US$150,000). Prior to tender invitations, the operator had to announce the tender schedule and companies to be invited. The Ministry's role was to ensure that qualified Norwegian companies were included on the bidders' list.

At the contract award stage, the operator was to inform the Ministry of their evaluation with the recommended supplier, price, country of origin and Norwegian content. The Norwegian content was calculated as value added in Norway both in manpower and monetary values. Ownership of the company was of less interest, what mattered was where the work was to be carried out, that is, in Norway or abroad. The role of the Ministry was to ensure that a Norwegian bidder was awarded the contract, when competitive in terms of price, quality, delivery time and service. If the Ministry was not convinced, consent could be withheld.

Information on upcoming tender invitations was also to be submitted to the Ministry of Petroleum and Energy twice a year by the operators. The Ministry collected the information and released it to all the Norwegian suppliers as a special service. Operators also had to file annual reports, with the intention of giving the Ministry additional information on previously reported contracts and also reporting (for statistical purposes) the value of services used, and finally, the Norwegian content (total value from each operator) of deliveries below NOK 1 million, as such contracts were not part of the current reporting.

The emphasis placed on Norwegian content made it an essential factor for all oil companies, and indeed, the Ministry used Norwegian content as one of the evaluation criteria when evaluating companies competing for new acreage. The Ministry's policy was to be transparent and predictable in respect to enforcement of the procurement policy. Moreover, as an observer in all license groups, the Ministry secured insight into all the operators' contracting activities. Good co-operation between the authorities, the operators and the contracting and supplier industry was essential in developing the high local content in Norway that at times exceeded 70 percent. On the conclusion of the agreement on the European Economic Area (EEA) in 1994 between Norway and the European Union, the procurement policy was discontinued. Presently, the Norwegian share of goods and services to the oil industry is about fifty percent. In hindsight, by 1994, the procurement policy had served its purpose and the Norwegian oil service industry was ready for exposure to international competition.

The establishment of the Norwegian state oil company, Statoil, in 1972 and the involvement of two Norwegian private companies, Norsk Hydro and Saga Petroleum in upstream oil and gas activities were intended to secure a central role for Norwegian industry. Increasing domestic competition and insight into international oil operations were useful for the Norwegian companies. The government actively established Statoil and encouraged the two private Norwegian oil companies, Norsk Hydro and Saga Petroleum through preferential licensing. International oil majors were placed in the role of technical assistants and joint teams were used to fast-track the Norwegian companies into fully-fledged operators. Consensus in Norway was that operatorship was necessary to learn the tools of the trade and to be able to meet foreign oil companies on an equal footing. At the outset, Statoil was financed by government loans and the central bank.[36] Financial independence was pivotal in securing a strong bargaining position with the international oil industry.

Licensing terms for prospective blocks in which the large fields were later discovered, stipulated that the operatorship was to be transferred to Statoil following completion of the field development. This brought Statoil quickly into a position as a significant operator and it is now the largest oil producer on the Norwegian Continental Shelf. Today, Norwegian oil companies (together with the state's direct financial interest, SDFI) accounts for approximately 84 percent of the production (as operator) and 70 percent of the reserves. Around 2000, Statoil had become the third largest seller of crude oil in the world market, after the national oil companies of Saudi Arabia and Iran. The national equity participation has also been around 70 percent. This has been regarded as an important element in achieving the goal of sharing the petroleum asset with the Norwegian population and developing a high local content in the procurement of goods and services.

Following its establishment in 1972, Statoil was allocated a minimum of 50 percent equity in all licenses, which brought the company into a position of significant influence. The explicit intention was to use Statoil as an instrument of industrial policy, to enhance the supplies of Norwegian goods and services.[37] In 1975, Statoil initiated the establishment of Norwegian Petroleum Consultants (NPC), where ten of the biggest Norwegian engineering companies formed the group as an engineering and project services contractor. Statoil encouraged the international engineering and management contractors to establish joint ventures with NPC and to perform their work in Norway, in particular for the development of the Statfjord B & C and Gullfaks platforms. To execute projects in Norway, i.e. engineering, procurement and construction management, in an economic and effective manner, the international contracting community also needed joint ventures with local partners. This meant that there was a call for Norwegian engineering contractors. The Joint Ventures had certain work principles on which the Norwegian engineering contractors established a basis for the transfer of technical know-how and experience. These were:

- Integrated Joint Venture organizations with formal deputies for key positions
- On-the-job training to gradually substitute foreign engineers with indigenous ones
- Focus on the use of young national engineers
- Procedures and job descriptions for all tasks and positions

While both ship-builders Aker and Kvaerner initially participated in NPC, they were also independent actors and grew to an important size. At its peak, there were three competing environments for field development in Norway. Later, Aker bought NPC, and Asea Brown Boveri (ABB) emerged as another player. ABB placed its headquarters for oil and gas in Norway and grew to international dimensions through acquisitions. Statoil and later Norsk Hydro have had key roles in developing high local contents through projects executed in Norway.

An important consequence of executing projects in Norway was that the tendering activities for contracting and procurement of goods and services also took place locally, in conformity with public policy. It often happened that the staff could provide information about capable local contractors and suppliers. Only if local suppliers did not have the capacity or the competence could tenders for contracts be open to international competition.

Knowledge Policy

The license terms for the international oil companies made it mandatory to transfer skills and competence to the Norwegian companies. Personnel from Statoil, Norsk Hydro and Saga participated initially in the oil majors' training courses and received on-the-job training schemes in their overseas operations. The oil majors recruited young Norwegian engineers and trained them overseas for a significant period, before they were taken home to "Norwegianize" their organizations. Some of these persons also ended up in the organizations of the Norwegian oil companies.

In the meantime, the skills gaps in Norway were filled through exchange programs with expatriate staff from the international oil companies. The immigration policy at that time admitted foreign experts to the Norwegian labor market if the domestic labor force could not satisfy the skills requirements. This policy applied to indigenous as well as foreign oil companies, oil service companies and engineering contractors. In order to better suit the basic training of the Norwegian workforce it was stipulated that the Norwegian language was to be used in the operating manuals and to label equipment. Experience from developing countries was that international oil companies at the time often were reluctant to train locals for the top jobs. In Norway, policy gave them no choice.

Today, both international oil companies and oil service companies employ local staff almost entirely and the trend is that these companies provide staff for their overseas headquarters and operating units. Norwegian human capital is being exported by international oil companies.

The transfer of technology and the cooperation in research and development have been among the most successful aspects of Norwegian petroleum policy. By compelling oil companies to transfer competence and to cooperate in the development of new technology, Norway could assume the role of a leader in international petroleum development. Within a relatively short period of time, Norwegian competence and technology for Norwegian conditions were developed. Competence strengthened Norway's bargaining position with the international oil industry. Technology development led to a significant cost reduction and an ensuing expansion of the resource base. During the 1990s, development costs per barrel of oil equivalent in new oil and gas fields on the Norwegian continental shelf most likely saw an annual reduction of 4–5 percent. The background was the strong research and development effort during the 1990s and the cooperation between oil companies, the supply industry and research institutions.

By experience, oil companies seem willing to finance research and development efforts, not only for their own immediate use. The motivation can be to ensure goodwill in host countries and to develop new solutions that can be applied elsewhere.

In Norway, the requirement to transfer competence and to cooperate in the development of new technology was introduced in the third licensing round in 1973. A practical result was that the international oil company Mobil, as the initial operator at the giant Statfjord field, systematically had to train Statoil to take over the task.

The fourth licensing round in 1979 introduced provisions for technology development cooperation between foreign oil companies and Norwegian research institutions. Cooperation agreements made the oil companies contribute funding, insight and expertise to develop technology in Norway. They were of three types. The first type, called "fifty percent agreement" required operators to conduct at least 50 percent of the research and development needed to develop an oil or gas field in Norway at Norwegian institutions. The second type of agreement used in the fourth and fifth licensing rounds required operators to conduct a specified research effort in advance of new licensing. The third type was a goodwill agreement, where the oil companies made an attempt to conduct as much petroleum-related research and development as possible in Norway, without any advance commitment as to the sum or volume of the effort. Initially, the research and development goodwill programs were based on dialogue with the industry and were developed to support its needs and priorities.

Technology cooperation soon grew to large volumes, especially the goodwill agreements. More than half went to engineering firms, the rest to Norwegian oil companies, maritime services, mechanical industry and research institutions. The value of the technology cooperation should be seen in the light of the lead times for technology breakthroughs, which are usually 10–15 years. From around 1980 to the mid-1990s, the Norwegian oil industry has experienced major advances in drilling and subsea technologies, the application of information technology, and hence, cost reduction. The formal technology cooperation was also discontinued in 1994.

One success story has been the development of subsea technology. Significant resources were allocated to subsea research and development in the early 1980s. Side effects from the research and development through industrial cooperation have been a larger degree of openness from oil companies towards the domestic industry, availability of oil company expertise, and coordination of research and development efforts of oil companies and suppliers. Important synergies were obtained by the joint development of technology.

In 2001, a wide cooperation venture was launched to establish a national strategy for research and development. The Oil and Gas in the 21st century (OG21) partnership seeks to meet the major value creation and environmental challenges relating to continued development of the Norwegian Continental Shelf and to strengthen the international competitiveness of Norwegian industry. An important goal of this initiative is to ensure more integrated and effective cooperation in the oil and gas cluster for research as well as for the demonstration and commercialization of technology. Attention is focused on achieving synergies throughout the entire research chain and on relationships between oil companies, suppliers and research institutes.

Given the most important challenges facing the petroleum sector, five main priority areas have been identified within the OG21 partnership:

- The environment
- Improving the recovery factor
- Deep water
- Industrial utilization of gas
- Marginal fields.

Another important instrument for promoting new technological solutions in the petroleum sector is the Demo 2000 collaboration, which focuses on three principal targets:

- Securing new field developments on the Norwegian Continental Shelf through innovative and cost-effective technology and execution models
- Increasing assurance that projects can be implemented in accordance with a budget and on schedule
- Developing new Norwegian industrial products for sale in a global market.

Innovative and cost-effective technology will be qualified for use through demonstrations (pilot projects). The aim is to ease the work of establishing new development projects, products and jobs.

The pilot projects involve close collaboration between suppliers, research institutes and oil companies. Such cooperation will in itself contribute to creating a forward-looking, market-focused expertise network. Funds for Demo 2000 are channeled through the Research Council of Norway. A major new petroleum research program called Petromaks has been launched in 2004, also under the coordination of the Research Council of Norway. On the basis of technology targets defined in OG21, Petromaks will seek to improve the utilization of producing fields and make new reserves more accessible. The key research tasks for the first phase of this new program will be:

- *Exploration:* developing methods for geophysical measurement, exploration and reservoir models and improved understanding of basin modeling;
- *Enhanced recovery in a broad perspective:* developing methods for simulated recovery, reservoir monitoring and management and drilling technology as well as new processes, methods and technology for gas sales.

In hindsight, the policy of transfer of knowledge expressed an ambition to use a natural resource-based, extractive and capital intensive industry, to strengthen the basis of a knowledge-based economy. In terms of economic dynamics, the oil industry is an extractive activity whose value added contains a large share of economic rent; the supply industry is a manufacturing activity whose value added essentially is based on labor productivity and management quality, and the research is an intellectual quality whose value added is essentially based on knowledge and inventiveness. Many oil producers remain stuck in resource extraction, living off the economic rent; some move into manufacturing, adding productivity to the income base; and a few, such as Norway, venture into research, adding knowledge. The challenge for Norway has been, and still is, to compensate the depletion of a finite resource with the creation of durable assets. Human capital or knowledge is considered more durable than real capital in manufacturing assets, as over the long run, investment in education outperforms practically any other investment.[38] From this perspective, Norway's oil policy has contained a knowledge strategy providing for a long-term income base, parallel to the investment fund. Rising exports of petroleum technology indicate a certain success.

Petroleum Revenue Management

In Norway, the quadrupling of the oil price in 1973–74 caused alarm. The Finance Ministry was concerned with overheating of the economy and advocated caution, meaning a more moderate rate of development of the oil industry as well as moderation in the use of oil revenues. Indeed, in a report to Parliament in early 1974, oil was to some extent associated with the problems of restructuring the economy. The message was that within a few years' time the country would have a huge financial surplus and the currency was duly revalued. At the same time, however, Norway's open economy was hit by the international recession in the wake of the oil price rise. Export industries and ship-owners were hit. Anticipating the financial surpluses, Parliament offered generous assistance, and also raised farmers' incomes to the level of industrial workers. The government embarked on a costly counter-cyclical economic policy that raised real incomes by 25 percent during three years, from 1974 to 1977, seriously compromising the competitiveness of non-petroleum activities. As a result of cost overruns and delays, oil revenues did not come as expected, and by the end of 1977, Norway had the highest debt ratio ever attained by any developed country, amounting to 50 percent of the gross national product, GNP. The currency was devalued and in 1979–80 the government called an economic emergency, freezing incomes and prices.

The crisis was of a short duration. Already by 1979 oil output was on stream as planned and oil prices were rising as a result of the Iranian Revolution. In the early 1980s, a new government gave priority to paying down the debt. Prospects were good, and spending could resume, until oil prices fell in 1986. The oil price decline provoked a change of government and a policy of economic austerity until the early 1990s. By that time, there was a wide consensus that oil revenues were a problem and should be handled with care, as advocated by the Finance Ministry in the early 1970s. The point is to use them indirectly.

Established by Parliament in 1990, the Petroleum Fund received its first true transfers in 1996. Its income represents the central government's net cash flow from petroleum activities, as well as the return on Fund investments. Proceeds from the privatization of Statoil also go into the Fund. Expenditures comprise an annual transfer to the Ministry of Finance corresponding to the amount of petroleum revenues applied in the fiscal budget to cover the non-oil deficit, plus the Fund's administrative costs. The Fund serves two purposes. It acts as a buffer which provides greater flexibility in economic policy should oil prices or activity in the mainland economy decline, and serves as an instrument for dealing with the financial challenges presented by an ageing population and the prospect of declining oil and gas volumes. In principle, the normal rate of return of the Fund, considered to be four percent, goes to the budget. In practice, there is some violation of the rule as Parliament tends to vote somewhat larger transfers in the annual budget transactions. By Norway's constitution, Parliament represents the supreme authority and cannot be bound. Consequently, there are no provisions safeguarding against sudden changes or opportunistic uses of the Fund.

Nevertheless, the consensus is that the Fund is a useful buffer against political temptations in a parliamentary democracy where politicians compete for votes promising outlays. With the present high oil prices (2005) the outlook is for a quick build-up of the Fund, which in a few years' time could have a value well exceeding Norway's GDP, providing long-term economic diversification. Norway also has large currency reserves.

The Finance Ministry has delegated responsibility for managing the Fund to the Central Bank. The assets are invested abroad, in shares and bonds. Shares currently make up about 40 percent of assets. From 1997 to 2004, the average annual return on the Fund's investment has been 5.76 percent—which measured in a basket of currencies and real return, corrected for inflation, has been 4.09 per cent. Management costs have been on the average 0.08 per cent, so that net real return has been 4.01 per cent on the average.

In addition to managing the Fund, the Central Bank also has to manage the exchange rate of the Norwegian krone, with the purpose of avoiding both an undue appreciation, compromising competitiveness, and an undue depreciation, fuelling inflation.

In hindsight, the establishment of the Fund has been a political challenge rather than an economic one. From the perspective of macroeconomic management, it was evident already at the time of the first oil price rise in 1973–74 that a mechanism should be established for the indirect use of oil and gas revenues. This would have been consistent with the policy of moderation and concerns for other activities. Unexpected turns in the international markets and opportunistic turns in Norwegian politics, causing the use of all revenues for almost two decades deferred its establishment. If, contrary to what really happened, a Fund had been established by 1975, it would have been several times its actual volume by 2004 and occasional periods of overheating and contraction, boom and bust, would have been considerably dampened. Significantly, the Norwegian economy was directly much less affected by the 1998–99 low oil prices than was the case in 1986–87.

The Significance of Norway's Experience

The historical background, especially the strong maritime tradition, makes any imitation of Norway's experience difficult. The hostile environment and the ensuing technical and organizational challenges are comparable only to a few regions, in or near the Arctic. Likewise, Norway's traditions of democracy and the accountability of its government have been decisive. Nevertheless, Norwegian policies could merit consideration in the Gulf countries, which are aiming to achieve economic diversification. It is, however, essential to stress that the Norwegian background is quite different from that of the Gulf countries, as is its industry.

1-The first major difference: contrasting technical challenges

In the Middle East, and especially in the Gulf countries, oil extraction takes place under comparatively lenient conditions, either onshore or in shallow waters offshore, generally from giant or super-giant fields. However, in the North Sea, conditions are comparatively hostile, often in deep waters, but always with harsh weather conditions, and in many cases from smaller or medium-sized prospects. The degree of adversity and less favorable geology mean that the geological, technical, environmental and financial risk is far higher in the North Sea than in the Middle East. The counterpart is greater technological and organizational challenges and stronger incentives for research and development.

At the outset, in the late 1960s and early 1970s, Norway at best appeared as a marginal oil and gas province, with a precarious break-even position. The quadrupling of oil prices in 1973–74 and the nationalization of the oil industry in many OPEC countries squarely put the North Sea in a central position on the international oil map. The enhanced interest in prospects strengthened the bargaining position of the Norwegian government. This did not reduce the technical challenges, but permitted a sustained and coordinated effort in petroleum-related research and development, drawing on domestic intellectual resources and the cooperation of the international oil industry. The results are to be found in subsea, processing, transportation and reservoir management technologies that are leading the world in some aspects. The counterpart is a comprehensive investment in human capital.

In the Arab Gulf states, the technical and geological challenges are less overwhelming, although there is a case for technology improvement, not least in areas such as in reservoir management to enhance the rate of recovery. Consequently, there is a good case for strengthening the effort in petroleum-related research and development in the Gulf states, but it should be focused on selective targets. The need is less to develop an entirely new petroleum technology than to improve aspects of the technology currently in use and eventually to become world leaders on some accounts. The requirement is a heavier focus on technical training and education in science. This matches local ambitions to develop a knowledge-based economy.[39]

2-The second major difference: contrasting economic background

Whereas the Gulf states, and much of the Middle East, until the early 1970s suffered from poverty and underdevelopment, Norway at that time was a developed and diversified economy, with a highly educated population enjoying full employment and a high standard of living. Consequently, in the 1970s, for the Gulf states, huge oil revenues appeared as an unmitigated blessing, but for Norway they also represented a risk and a menace to the established industries. For the Gulf states, oil revenues represented an enormous supplement to the economy and the opportunity to leap in a few years' time phases of development that in other conditions can take generations—not the least in matters of education and health. For Norway, by contrast, oil revenues not only represented a supplement to the economic base, but also a substitute for established activities, so that the net gain would be less. Moreover, the oil industry, by raising the domestic cost level, has appeared as a threat to employment, as it could easily destroy more jobs than it creates. Against this backdrop, there has been a constant pressure to diversify for the sake of employment, also through the oil services and supply industries, and in this way enhance the value added in relation to oil and natural gas.

In the Gulf states, there is a good case for engaging more systematically in the petroleum procurement industries, but this requires skilled labor at competitive costs and competent management. Even with a large oil industry, the Gulf states have historically imported most equipment, components and parts. The overhaul and subsequent expansion of the Gulf oil industry mean heavy investment for many years and corresponding procurement needs. The risk is that short-term convenience will direct Gulf oil industry procurement contracts to friendly foreign firms. Instead, the procurement program should be handled by locals, benefiting from the occasion to develop a Gulf oil supply and services industry, which eventually could serve markets in neighboring oil-producing countries. This could be done in cooperation with foreign investors, oil and manufacturing companies to benefit the trade balance, employment and competence. The employment situation varies. Some Gulf States rely overwhelmingly on foreign labor for manual tasks and domestic labor is extremely costly. Other Gulf states, not the least Saudi Arabia, have an employment problem, especially among youth, and a pressing need to create more jobs. Insofar as the domestic oil industry needs maintenance, renewal and eventually capacity expansion, there is a domestic market for goods and services that presently is not much by local suppliers. Again, the need is to be focused and selective, at least in an initial phase, and to target some points in the upstream petroleum value chain. A relevant ambition would be exports of some goods and services for the international oil industry. Again, the requirement is the appropriate training of labor and managers, as well as willing investors.

3-The third major difference: political economy

Although Norway, like the Gulf states, is a monarchy, it is indeed of another character. In Norway, the role of the state is traditionally that of a redistributor of money, levying taxes on incomes, turnover and property, whose sums are reallocated by a democratically elected parliament. Although oil revenues are high, as Norway in oil and natural gas extracts about a barrel a person a day, they have not altered this basic function of the state. Indeed, it is by maintaining taxation on the citizens that it has been possible to build up the Fund and use the petroleum revenues indirectly.

In the Gulf, the role of the state is essentially to distribute petroleum revenues, avoid taxing the citizens, and hence preclude the need for democratic control.[40] The risk is a vicious circle. In the absence of any handover of power to democratically elected bodies, taxation of the citizens is deferred, thereby prolonging and intensifying the dependence on petroleum revenues and the ensuing risk. Among the Gulf states, medium-term economic prospects differ profoundly, and in some cases a prolonged dependence on petroleum revenues does not seem to entail high economic and social risks. In other cases, the outlook is quite different, not the least because of high population growth. Here, the pressure for diversifying the income base and employment opportunities is likely to intensify. The crunch could be the willingness to tax the citizens in return for a more formal political participation. Without the state securing a broader income base, there would be scant hope for using oil revenues more indirectly, and hence, an increasing risk of discontinuities in economic and social development.

In Norway, both the state and the market have had a role in the policy of economic diversification in relation to petroleum. One key lesson is that market forces alone never would have accomplished the transfer of competence, the successful establishment of supply and service industries, or the Fund. If left to market forces, in the economy as in politics, chances are that oil companies would have shown a preference for foreigners with a proven competence, established foreign suppliers of goods and services, often with a long-term symbiotic relationship with the oil industry, and politicians would have competed shamelessly for the maximum use of petroleum revenues, combining promises of tax breaks with generous pledges for public expenditure, as in some other oil exporting countries. The outcome would most probably have been conflict, strife and political discontinuities to the detriment of all parties involved.[41] As mentioned, the market is less able to meet long-term structural and strategic challenges.[42]

Against these risks, a long-term state strategy has worked out surprisingly well, based on a broad political consensus. The consensus has contributed to a robust petroleum policy, with a remarkable continuity through shifting governments. The consensus, in turn, is based on a long history of representative, transparent and accountable government. For example, Norway's Petroleum Fund by law has to publish annual reports on its operations, and they are under the scrutiny of the Comptroller General, appointed by Parliament. This is also the case for other public revenues and expenditure.

The other key lesson is that the state alone never could have accomplished its objectives. It needed the cooperation of a competent and willing private sector. This is also a question of demarcation, of agreeing on sensible borders between the private and public sector, based on a reasonable assessment of comparative advantages and economies of scale.[43] The tradition of shipping, shipbuilding and mechanical industry, all in private hands, evidently was of great help. Even if the primary petroleum industry in Norway still has a high degree of state ownership, there has never been any question of nationalizing the oil supply and service industries. Indeed, the role of the state was essentially to facilitate petroleum business opportunities for the private sector. The state in its licensing policy honored foreign oil companies that gave a priority to Norwegian goods and services, measuring the value and the domestic content percentage, but refrained from specifying Norwegian goods and services that should be preferred. This policy set the framework for a preferential Norwegian petroleum supply and services market, but with competition and flexibility and no central planning.

Loose coordination between ministries ensured a high degree of arbitrariness, indicating practical limits to state action, at least as concerns implementation and details.[44] The Ministry of Petroleum and Energy would grant licenses, the Ministry of Industry would evaluate industrial projects and applied research, and the Ministry of Education would supervise basic research and training. The government departments provided incentives and it was up to the partners to find and develop the projects. This *ad hoc* approach meant that many projects in hindsight were of dubious value, but the cumulative effect was a significant build-up of competence and insight that was essential for the technological progress and ensuing cost reduction in the offshore Norwegian petroleum industry, which was beneficial to the oil companies, the Norwegian supply industry and the Norwegian government, the tax collector, alike. In hindsight, however, the preferential policy went too far, as around 1990 at least four fifths of the new prospect content was domestic. The advantages were jobs and profits in Norway, but the disadvantages were too few and too weak impulses from abroad, too much dependence on the petroleum industry for Norwegian manufacturing and too little exports to markets in other oil producing countries.

Norway's experience shows clearly that access to petroleum resources can be used as a bargaining tool, especially to obtain benefits that are of a common interest, strengthening the mercantilist state. Today, in a situation of a tight oil market, this evidently also applies to the Gulf states. Their major advantage is the availability of huge proven oil reserves at low cost, which enhances attraction for the oil industry. Their major disadvantage is the absence of a manufacturing base and a correspondingly skilled labor force, as well as the paucity of technological and scientific research. Consequently, it would be futile to emulate the Norwegian experience in all aspects.

In relation to foreign oil companies as well as foreign governments, the Gulf states have a strong bargaining position as major sellers of crude. There is little economic need to invite foreign oil companies as upstream investors because the Gulf States have sufficient financial resources. Concerns about security of supply give long-term crude contracts an additional value above the market price, which can be used for negotiating other advantages. The Gulf states might consider long-term comprehensive deals with crude sales progressively supplemented by product sales. They might also consider payment as knowledge transfer, human capital development and industrial investment in addition to money.

Using long-term oil supply contracts as a bargaining tool to obtain benefits other than monetary payment would mean an effort to maximize the value of oil beyond the immediate market price. In this effort, the Gulf states and other oil exporters are helped by the restructuring of international oil trading: that is taking place through comprehensive bilateral agreements, as between Iran and China, and India. This development seems to signal a trend toward neo-mercantilism that conflicts with the liberal globalization prevalent since the early 1980s. The reason is simply that security of supply again is an issue; many importers realize that the most expensive barrel of oil is the one that is not delivered, whether due to disruption of current supplies or the non-availability of incremental supplies. China's strategy is evidently to conclude comprehensive deals involving investment capital, oil and natural gas supplies, and exports of manufactured products. India is emulating and competing with China. Iran is following similar strategies, with major deals with China and India. China is successfully pursuing this strategy in relation to other potential partners, such as Venezuela and Sudan. The question now is whether Japan and Korea will follow suit, and what Europe will do eventually.

However, the strong commercial tradition in the Gulf states could help, especially in cooperation with foreign partners. Access to petroleum can be used as a bargaining chip. Incentives do not work with exclusion. With long-term supply contracts, it would be possible not only to require foreign oil companies to train locals and transfer knowledge, but also to help develop some local supply and service industries and even bring them along to markets in other oil provinces. In this respect, the Norwegian experience could provide a useful example, positive as well as negative. The merchant class of the Gulf sates could provide the local partners for the international oil industry, and subsequently help in diversifying the economy by establishing a more independent income base. With the prospect of a long-term access to oil, foreign oil companies would have an evident self-interest in training locals to develop a pool of skilled local labor.

Regional cooperation is another clue. The diversity of the Gulf states contributes to the complementary nature of their economies and labor markets. For instance, Saudi Arabia has a youth unemployment problem unlike the smaller Gulf states. Regional cooperation aiming at a common economic diversification would be able to draw on larger markets, more human resources and more capital than an effort by each of the Gulf states. Policies to build up a common base of petroleum supply and service industries would run into rivalries and jealousies among the Gulf states, but the benefit would be a stronger bargaining position with the international oil industry. The alternative could be parallel efforts and a fragmented bargaining position.

The first priority could be to diversify within petroleum by developing a local supply industry. Prospects for the oil market indicate a need to maintain and eventually expand capacity. Today, there is little local content in the deliveries. Developing a local or regional oil service industry could mean diversifying income sources and at the same time creating jobs. The need and the potential in this direction are evidently greater in Saudi Arabia than in Kuwait, Qatar or the United Arab Emirates.

The second priority could be to develop competence in relation to petroleum. This would be most successful in cooperation with the international oil industry, which would mean opening the industry to foreign operators. This would be politically controversial in some countries, but relying on a single, nationalized operator unfortunately means isolation from competence developing elsewhere.

The third priority would be to develop a Fund, aiming at using petroleum revenues indirectly. This is already the case in Kuwait and Abu Dhabi, but should be considered in Saudi Arabia in the light of the currently high oil prices that might not last forever. Indeed, the combination of demand and supply factors, including the political risk widely perceived in the oil market, which is pushing oil prices above $40 per barrel is unlikely to last. Consequently, a wise strategy would be not to adapt expenditure to temporarily and unexpectedly high oil revenues, but rather to put unexpected income aside in a stabilization fund to cushion the domestic economy and support diversification.

Conclusion

In the case of Norway, an appropriate question is to what extent the various protectionist policies in relation to the petroleum industry, concerning local content, knowledge transfer and industrial employment, succeeded in reducing dependence on the petroleum industry, although causing the country to suffer from a bout of "Dutch disease" in disguise—a resource curse intensified by institutions. Indeed, the local content policy had made parts of Norwegian manufacturing exceedingly dependent on the local offshore market. The knowledge transfer policy had concentrated the country's research efforts on petroleum-related matters. Significantly, as these policies of micro-economic intervention were abandoned in the mid-1990s, there was a general understanding that they had contributed to high costs in the petroleum industry as in other parts of the economy.

Since the mid-1990s, Norwegian policy has been keen to accumulate the petroleum revenues in an investment fund, the intention being to avoid harmful domestic inflationary pressures due to the expenditure of economic rent. So far, this policy of macro-economic management has been reasonably successful, but the risk is that huge surpluses on the current account balance will drive the exchange rate up, raising the domestic cost level and compromising the competitiveness of other industries, the resource curse materializing through a financial endowment. In that case, the benefit for the country of keeping oil and natural gas in the ground might exceed the value of financial surpluses and the question in relation to the resource would be to extract or not to extract.[45] If this were to be the case, the resource curse hypothesis of Sachs and Warner would be verified. Otherwise the contrary hypothesis of Halvor Mehlum, Karl Moene and Ragnar Torvik would seem plausible.

The latter hypothesis is based on an assumed distinction between producer-friendly institutions that make rent-seeking from resource extraction and production into complementary activities, and grabber-friendly institutions, which make them competing activities. Evidence from Norway is mixed on this account. At the beginning of its petroleum activities, Norway clearly benefited from the rule of law, a well-functioning civil service and a low level of corruption, meaning producer-friendly institutions. This framework attracted entrepreneurial talent to petroleum-related activities. The swift and successful implementation of protectionist measures signified, however, that the institutions were also grabber-friendly, underlined by the early use of anticipated petroleum revenues in the 1970s to support farmers, ship-owners and manufacturing in general. By the mid 1990s, the oversized oil supply and service industry indicated that Norway's institutional framework had some grabber-friendly aspects. One test is the export performance of the oil supply and service industry, which at the time of writing is growing, indicating international competitiveness and that Norway's institutions are producer-friendly.

The contrary hypothesis put forth by Mehlum, Moene and Torvik seems to assume that political institutions are neutral in relation to their economic base, an assumption hardly supported by evidence.[46] Large resource revenues make the government more independent from a productive income base and from taxes levied on civil society, usually leading to a consolidation of power, in developing countries often strengthening dictatorships. Oil revenues are the foundation of the autocratic rentier states of the Middle East, making oil wealth a substitute for democracy and an impediment to institutional development.[47] In industrial countries with established democracies, such as the Netherlands or Norway, large resource revenues to some extent decouple expenditure from productivity, enlarging the freedom of action in economic policy, making the political system a more tempting battle ground for interest groups, and encouraging arbitrariness and rent-seeking through selective measures. Norway is a case in point. The country has, so far, avoided the worst excesses of the resource curse, but huge petroleum revenues do carry substantial risks.[48]

Another question is to what extent the reliance on market forces, with a less interventionist state in petroleum matters and a stronger position for foreign companies, would have produced comparable results in areas that have mattered to the Norwegian public, namely the integration of the petroleum industry into Norwegian society through jobs, deliveries and national control. This is a complex issue, not limited to monetary costs and benefits, but also to qualitative aspects of Norwegian society that are difficult to quantify, let alone monetize. In the early 1970s, the emerging petroleum activities and related policies were the subject of intensive political debate in Norway, with the government presenting alternative policies and risks, which has been described as exemplary for democratic deliberation of resource policy.[49]

Section 3

OIL MARKET IMPACT

6

World Oil Prices: Impact on Economic Growth

Mehdi Varzi

In view of the current uncertainties in the global economic environment and the political storms affecting the oil market in recent months, an international gathering of energy experts and observers such as this ECSSR conference assumes particular importance.

Politics versus Economics

When a subject such as oil and the world economy is under discussion, there are a few simple truths that are sometimes ignored or forgotten. One question that often arises is why oil remains such an intensely political subject. To answer this question, it is necessary to recall the following realities:

- Oil has played a key role in the industrialization process of the world. Governments have, therefore, tended to exert influence over their domestic oil industries and major powers have tried to gain control of oil reserves abroad.
- The really large oil reservoirs have been found in a relatively small number of countries, concentrated overwhelmingly in the Middle East.
- Oil plays a vital role as a source of fuel for the modern military machine, whether on land, at sea or in the air. Imperialism in the twentieth century was in part an attempt by outside powers to gain

access to large oil reserves, although there were, of course, many other reasons. In short, the impact of oil goes far beyond the realm of pure economics.

- Oil accounts for up to 10% of international trade. No other commodity can match such a percentage figure. This trade is vital for the continued health and prosperity of economies worldwide.
- Some 40–50% of that trade is in the hands of OPEC, all of whose members depend heavily on oil revenues. Oil prices have a direct impact on their political fortunes, and they cannot afford to leave oil prices solely to the whims of the spot and futures markets.
- Oil has in the past had a major impact on currency movements and therefore, to an extent on inflation, with all its political connotations.
- Oil prices have a psychological effect on the prices of other commodities.

However, whereas the above-mentioned factors point to the prospect of continued interaction between the politics and economics of oil, it is almost impossible to quantify accurately at any single period in time what percentage of the oil price is based on economic as opposed to political fundamentals. So, it is important to clarify one point from the very beginning. If somebody had asked me in September 2004 whether oil prices would reach around $50 per barrel within 12 months, I would have replied in the negative. Many experts at this gathering have an even longer experience in the oil market than I have—and I began my career in the oil industry some 36 years ago. However, what is certain is that regardless of the years of experience or the extent of expertise one can claim, surprises will always occur. So it may be worth glancing briefly at the past behavior of the global oil market to consider whether there are lessons to be learnt for the future.

Above all, what are the key similarities and differences with the major oil price spikes of 1973 and 1979? It may be noted that I have avoided the word "shock" since, for OPEC producers, high prices are a windfall whereas low prices are a "shock." For producers the real shock was 1986 when prices fell to $10 per barrel. I have also omitted reference to the price spike of 1990–91, since it lasted barely six months and was followed by years of much lower oil prices. In short, in my view its economic impact was minimal.

Oil Price Spikes of 1973, 1979 and 2004: The Similarities

- *Global demand*

Assuming demand growth of around 2.6 million barrels per day (mbpd) in 2004, the three years from 2002–2004 would have witnessed total demand growth of around 5.2 mbpd. Over the past thirty years, there have been only two three-year periods when this strong rate of growth has been matched—first, between 1986 and 1988 following the oil price collapse and second, between 1976 and 1978, just ahead of the second oil price spike. So the last few years have seen a very high rate of sustained demand growth.

However, over the past fifty years, the most spectacular period of sustained demand growth occurred in the 1960s—a period of very low oil prices when even in many OECD economies, demand was growing at the same rate as, or faster than, GDP. Incidentally, in my personal view, given the enormous increase in demand in the decade prior to 1973, the tripling of oil prices was entirely justified. The Arab oil embargo may have provided the trigger, but prices would have risen sharply in the 1970s even if the embargo had not taken place. Between 1961 and 1971, the demand for OPEC crude tripled from 11 mbpd to 31 million barrels per day (mbpd). This may explain why the world economy recovered more rapidly from the price hike of 1973 than in the aftermath of the hike of 1979–81.

- *Demand for OPEC crude*

Is it just a coincidence or is there a lesson for us all? In 1973, 1979 and in 2004, price spikes have coincided with the demand for OPEC crude reaching the 30 mbpd level (see Table 6.1). At this level, OPEC is viewed as producing at virtually its maximum rate of capacity. Suddenly, the world has begun to discuss shortages of supply. Perceived or actual supply disruptions in key oil-exporting countries assume front page significance. Political events, especially in the Middle East, become daily news headlines. Thus, at such high levels of demand for OPEC crude, geopolitics tends to supersede economics.

Table 6.1

World Crude Supply/Demand (including NGLs) 1974–2004

Demand	*1974*	*1979*	*2004**
Total demand	**56.6**	**65.5**	**82.3**
Supply	*1974*	*1979*	*2004**
Total non-OPEC	**27.6**	**34.4**	**48.0**
Total OPEC of which:	**31.0**	**31.5**	**32.9**
Crude	30.6	30.6	29.2
NGLs	0.4	0.9	3.7
Processing gains	0.4	0.9	1.8
Total supply	**59.0**	**66.8**	**82.7**
Implied stock change	2.4	1.3	0.4
OPEC's market share (%)	54.8	48.1	40.0

Note: *Forecast: Varzi Energy.
Sources: IEA, OPEC.

In my personal view, OPEC has always had difficulty in raising production capacity substantially beyond 30 mbpd. In the early 1970s, certain member countries were more concerned with gaining control over their natural resources rather than boosting capacity. In 1979–1981, the revolution in Iran and the Iran–Iraq War resulted in a massive destruction of capacity in Iran as well as Iraq. Falling global demand in the wake of high oil prices and recession added to the loss of capacity as many OPEC members cut their exploration and development budgets. Today, while demand is resilient, some OPEC members question the need to raise capacity substantially, arguing that too aggressive a rise could result in a glut of oil in the market and another round of low oil prices several years from now.

However, there is another pressing problem—the challenge of reservoir depletion. This issue is well understood in many non-OPEC economies where the private sector plays a preponderant role in upstream activities and has to report on the matter in its annual report and accounts. However, upstream activities in most OPEC countries remain for the most part under state control. As such, facts and figures are much harder to extract. As the years go by, many of OPEC's major reservoirs, including the so-called super giants in the Middle East, are facing declining reservoir pressures. In my estimate, OPEC as a whole needs to replace around 1.3 mbpd of its annual capacity just to maintain capacity at prevailing levels. In short, whether or not the world at large considers these matters, such a depletion rate raises more doubts about whether OPEC will or can raise capacity to meet future world oil demand growth and how it will do so.

Finally, there seems to be a trend among some OPEC members to move away from co-operation with the major international oil companies. In my view, this is unfortunate, especially at a time of tight global upstream capacity. It is becoming increasingly clear that some OPEC members such as Venezuela and Iran need external expertise – technological and managerial – to boost their production to higher levels.

OPEC's belated measures to raise production capacity and unwillingness among some members to welcome foreign private capital into upstream activities are a genuine source of concern. Such an attitude tends to heighten geopolitical concerns about OPEC's ability and/or willingness to meet rising world demand. Thus today, even though there is no evidence of a real shortage in global crude oil stocks, a combination of tight downstream capacity and geopolitical concerns has kept crude oil prices way above the previously anticipated levels.

- *The Role of non-OPEC Producers*

Prior to the 1973 oil price spike, the world did not really distinguish non-OPEC producers from OPEC. In fact, OPEC was hardly recognized as a major player even though the group as a whole accounted for around 60% of global supply. This distinction between OPEC and non-OPEC really began in the 1970s as politics assumed ever-growing importance in the oil market and economists began to fret about an "impending global supply shortage." One might recall the famous report by the Club of Rome in 1972, which warned against an unduly rapid use of the world's mineral resources, especially oil. Over thirty years later, the world is at a juncture where global oil supply has kept pace with demand even though demand has risen by some 70% (see Table 6.2). Moreover, over the same period, non-OPEC supply has virtually doubled from 25 to 47 mbpd.

Table 6.2

Major Rises in non-OPEC Crude Supply to 2010

(including NGLs, thousand barrels per day)

	2003	*2004*	*2005*	*2006*	*2007*	*2008*	*2009*	*2010*
Russia	8460	9160	9500	9900	10,250	10,450	10,600	10,950
Angola	880	990	1190	1600	1800	1900	2000	2100
Brazil*	1770	1790	2000	2080	2150	2300	2480	2500
Kazakhstan	1026	1185	1270	1400	1475	1650	1850	2050
Azerbaijan	310	310	395	575	600	900	960	1100
Sudan	270	330	400	520	570	620	670	720
Canada	3000	3085	3075	3125	3175	3225	3275	3325
TOTAL	**15,716**	**16,850**	**17,830**	**19,200**	**20,020**	**21,045**	**21,835**	**22,745**

Notes: *Includes alcohol production.

Fears of a long term shortage of world oil supply are overdone. Supply has always been essentially a function of price. However, with demand so resilient to high oil prices, additional supply can only come if prices remain substantially above the levels of the 1980s and 1990s. Non-OPEC output is likely to carry on growing for the foreseeable future, but is unlikely to match annual demand growth.

Source: Varzi Energy.

Incidentally, I do not agree with the view that world oil supply or even non-OPEC supply is about to peak. This forecasting of imminent production peaks goes back a long way—indeed as far back as the 1880s when oil magnate John D. Rockefeller, fearing an impending peaking of US supply, sold his shares in his company. However, with more non-OPEC supply likely to come from non-conventional sources and difficult-to-access areas such as deep water locations, the cost of future non-OPEC supply is almost certain to rise (see Table 6.3).

Table 6.3
World Proven Crude Oil Reserves and Production Since 1960

	1960	*1970*	*1980*	*1990*	*2000*	*2003*
Reserves (billion barrels)	302	611	649	1,002	1,016	1,148
Production (mbpd)	21.1	45.1	62.8	66.8	76.2	79.9
Reserve life (years)	39.2	37.1	28.3	41.1	36.5	39.4

Note: Contrary to fears of reserve and supply shortages expressed in the 1970s, world reserves remain plentiful despite a huge rise in demand. However, future crude oil discoveries are likely to be more expensive to develop than in the past.

Source: *International Petroleum Encyclopaedia, Oil & Gas Journal*, BP and Dresdner Kleinwort Wasserstein.

In passing, I would comment briefly on the view of well known experts such as Hubbert and Campbell that oil supply is dictated by field geology alone. The above-named discount the importance of factors such as price, investment levels, technological advances and petroleum legislation. In my personal view, the world is nowhere near exhausting its physical oil resources. The current level of oil prices, as in 1973 and 1979, is due to a number of factors—both economic and political. However, certain developments are clearly occurring, which have potential long-term implications. The fact that prices are remaining stubbornly high not just today but in the long-term paper market sends a message—that the era of cheap oil prices is probably over. As in the past, high prices will extract higher global supply and undermine demand growth, resulting in a new equilibrium. Over the longer term, oil prices must reach a level that will allow other energy sources to help oil meet the inexorable rise in future global energy demand in a gradual manner over time. However, I believe that global oil supply is still many years away from actually peaking.

Comparing 1973, 1979 and 2004: The Differences

- *Tightening of the global supply chain*

The years 1973, 1979 and 2004 are remarkably similar in this respect. The entire chain of supply, from the drilling industry to shipping, refining and specialized labor looks tight today. Much of the tightness can be explained by the low oil price mentality of the past and, until recently, by the oil industry's doubts about the sustainability of an oil price above the range of $20–25 per barrel.

- *The rise of paper markets*

Such markets did not exist in 1973 or 1979. They have helped to create much greater flexibility and transparency globally, although one by-product is the appearance of much greater price volatility. Today, it is not uncommon to see daily paper trading equivalent to four or five times the physical level of daily demand or even more. In the consuming countries, more and more oil markets have been liberalized. This trend is spreading to emerging markets as well. Today, oil prices are closer to reflecting the market realities than in the past. This should make for a faster adjustment in consumer behavior and, hopefully, prevent a damaging sustained long-term rise in oil prices to unsustainable levels.

- *Declining oil intensity*

The role of oil as a share of GDP has declined markedly over the past three decades (see Figure 6.1). If one takes 1973 as the base, today the oil GDP ratio has fallen below 0.6:1 globally. This goes a long way towards explaining why the global economy is likely to continue growing in 2005 and possibly, albeit at a somewhat slower rate, even if prices average as high as $40–50 per barrel in 2006.

Figure 6.1
Ratio of World Oil/GDP since 1973

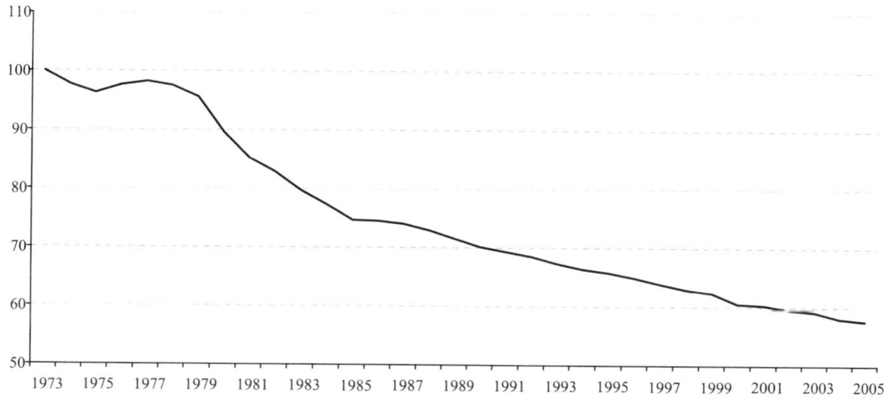

Note: Continued falls in the world oil/GDP ratio will tend to limit annual incremental demand growth.

Sources: Organization for Economic Cooperation and Development (OECD), Dresdner Kleinwort Wasserstein (DrKW) estimates.

- *The accumulation of government-held strategic stocks*

This development was a direct result of the first oil price spike in 1973. Today, the OECD alone holds well over 1 billion barrels and there are signs that some of the major emerging market economies like China and India will shortly follow suit. In theory at least, such a huge stockpile could be used pretty effectively in the event of a major global supply disruption, thereby mitigating a future oil price spike. Of course, in practice, what matters is that such stocks are employed in a credible and timely fashion as and when the need arises.

- *Consumer taxes*

Consuming governments, especially in the OECD, often earn more annually in taxes on oil products than OPEC members earn from their oil exports. In countries where product taxes are high, as in the key European economies, demand has risen at a far lower rate than in previous decades. Moreover, given that the crude oil price is a far smaller component of the total retail price, rising crude prices have a much smaller impact than in the past.

Outlook for the Global Economy

Global economic sentiment appears to have worsened appreciably over the past couple of months. This has been due to a combination of factors, including continued concern about global terrorism, rising oil prices and the loss of momentum in key economies, especially in the OECD (see Figure 6.2). Consensus forecasts for OECD growth have become more bearish in recent months (see Figure 6.3).

Figure 6.2
OECD Lead Indicators: 6-Month Annualized Rates of Growth % (2001–2004)

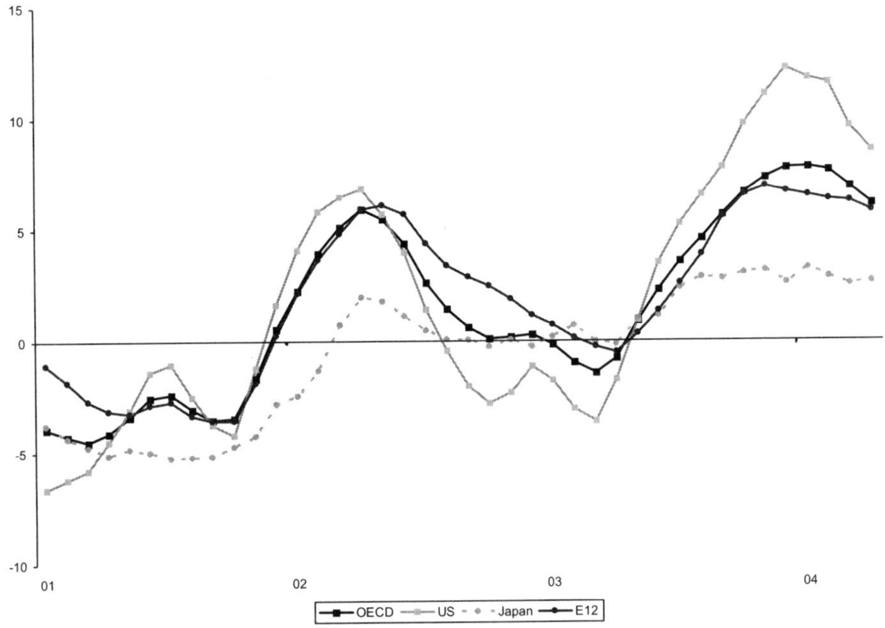

Note: Overall OECD appears to have reached a short-term peak.

Source: Dresdner Kleinwort Wasserstein (DrKW).

Figure 6.3
Consensus Growth Forecasts for OECD GDP

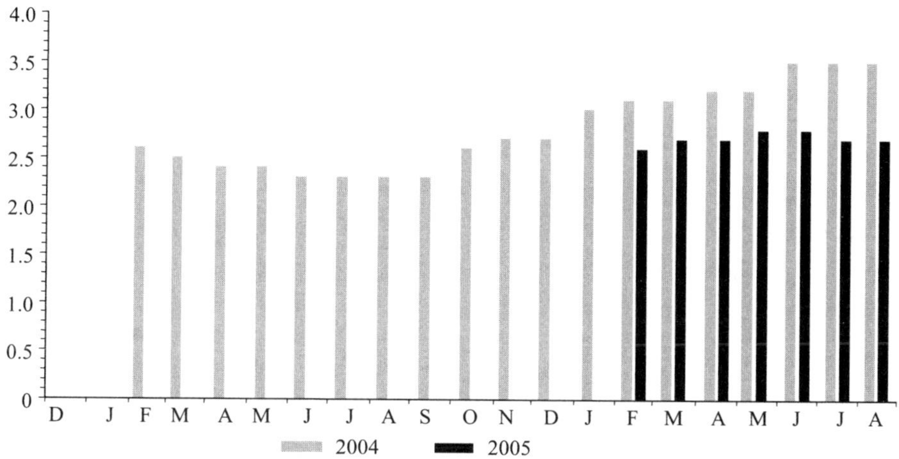

Source: DrKW Macro Research and Consensus Economics.

However, according to the latest forecast of Dresdner Kleinwort Wasserstein (DrKW), the world economy would continue to grow in 2005, albeit at a slower pace (see Table 6.4).

Table 6.4
World Economic Growth (Real GDP, Annual Growth %)

Annual Growth	*1997*	*1998*	*1999*	*2000*	*2001*	*2002*	*2003*	*2004*	*2005*	*2006*
World	**3.7**	**2.2**	**3.3**	**4.0**	**1.3**	**2.1**	**3.0**	**4.0**	**3.2**	**3.1**
of which:										
Asia (ex. Japan)	5.4	1.4	7.1	7.5	3.7	5.9	6.3	7.6	7.4	7.0
LatAm	4.4	2.3	-0.1	4.1	0.0	-0.8	2.0	5.7	4.4	4.1
Euro area	2.3	2.8	2.6	3.4	1.5	0.9	0.5	1.8	1.5	2.0
Japan	2.9	-1.1	0.8	2.2	-0.3	0.1	2.7	2.7	0.8	1.4
UK	3.0	2.6	2.3	3.0	2.3	1.7	2.2	3.1	2.1	2.3
US	4.4	4.4	4.1	3.8	0.3	2.4	3.0	4.4	3.4	2.5

Notes: Global GDP growth in 2004 was surprisingly high in the face of very strong oil prices.
Global GDP is likely to slow but remain at a respectable 3% or more in 2005/2006, despite the prospect of continued strong oil price performance.
Chinese GDP growth in 2005 is forecast at 10% ? 1% higher than in 2004. Oil demand should continue rising strongly though at a slower rate than over the past two years.

Sources: International Monetary Fund (IMF), World Bank and Dresdner Kleinwort Wasserstein (DrKW) estimates.

At first sight, such an assessment may seem surprising given the recent rise in oil prices. However, with prices still way below the real highs of 1979–81 and with the world less dependent on oil as a source of energy than previously, the global economy is likely to prove relatively more resilient than in past decades (see Figure 6.4). Again taking the latest DrKW assessment, a sustained oil price of $40–50 per barrel is likely to reduce US and Eurozone GDP growth each by 0.7% over two years. However, Japan is likely to see growth cut by as much as 1.5%, given its exposure to the Asian economies, which are generally more vulnerable to higher oil prices. Incidentally, DrKW forecasts that oil at $40–50 per barrel would reduce Chinese growth by 1.2%.

Figure 6.4

Real and Nominal Oil Prices (dollars per barrel)

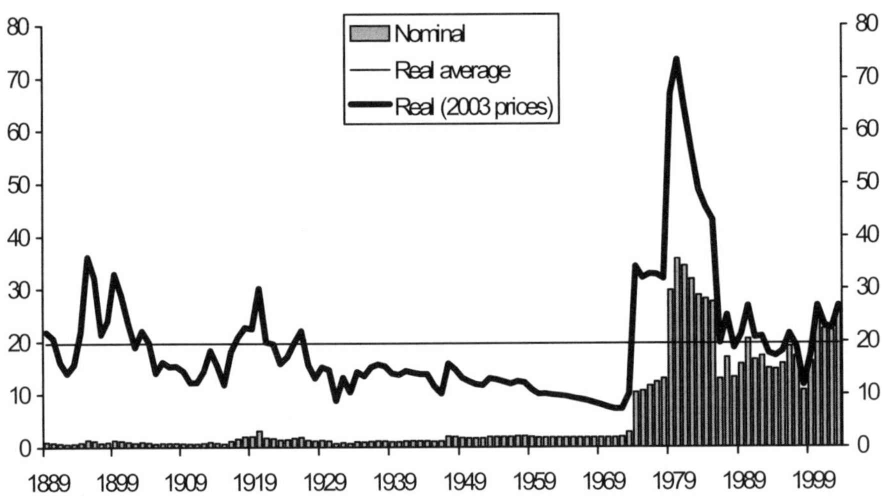

Note: Oil prices are way below the levels of 1979–81 in real terms.

Source: Dresdner Kleinwort Wasserstein (DrKW).

One danger is that higher oil prices could feed through into higher inflationary expectations, which might in turn result in central banks raising interest rates further than currently anticipated. Fortunately, there is currently little sign of such a development although inflation numbers are bound to be watched closely in the near future. In the Eurozone area and Japan in particular, domestic demand has remained very weak and economic growth has for some time been painfully slow—certainly too slow for manufacturers to pass on the full cost of higher energy bills to consumers. In Europe also, consumers have been sheltered by the dollar's weakness against the Euro, although it is unclear whether or not there will be a reversal in this trend in the coming months. An additional factor in Japan's favor is that the country is one of the most energy-efficient countries in the world, consuming only half of what the United States consumes on a per capita basis.

Potentially Positive Effects of Higher Oil Prices

Higher oil prices generally tend to have a negative effect on the global economy although, as mentioned above, prices have not yet gone high enough to trigger a global recession. What is less frequently mentioned is that higher prices also carry some potential longer-term benefits, two of which are outlined below:

1-Higher Exploration Budgets

Those who project the imminent peaking of world oil supply are too alarmist. The supply of commodities has traditionally tended to be a function of price. The oil market is no exception to this rule even though ultimately reserves are finite. Upstream technology has progressed remarkably over the past twenty years so that today oil is being discovered in commercial quantities in regions that were formerly beyond the industry's capability. Higher oil prices are triggering higher exploration and development spending, with more and more companies prepared to go into deep water areas (defined as below 1000 meters). After plunging to a 25-year average annual low of only 1457 active rigs worldwide in 1999, the number had recovered to 2395 rigs by 2004 and in the first two months of 2005, it averaged 2705 (see Figure 6.5). No doubt, higher prices will make it more attractive to explore for and develop oil in more difficult areas of the globe as well as increase the attractions of non-conventional oil such as the tar sands of Canada and the heavy oil of Venezuela.

Figure 6.5
Worldwide Rig Count (1976 to February 2005)

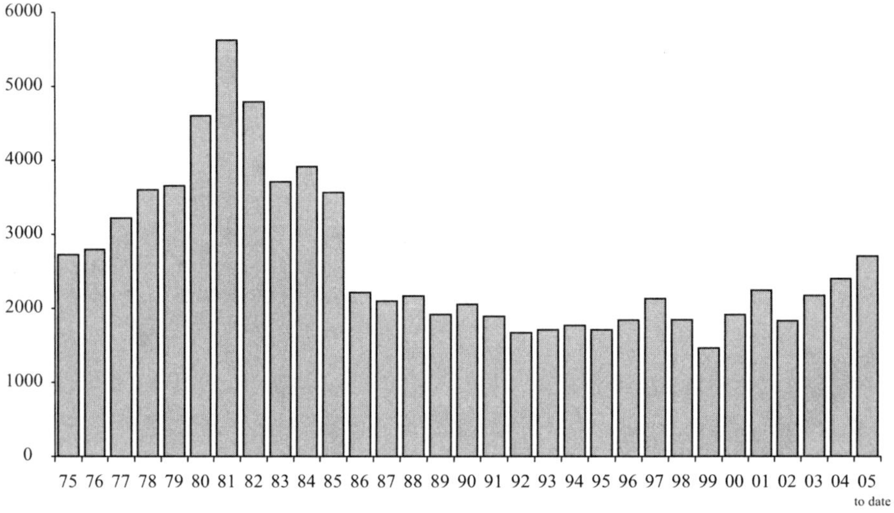

Source: Baker Hughes

2-Greater focus on energy conservation

Aside from recession, one main reason behind the severe fall in global oil demand between 1979 and 1983 was the wave of oil conservation measures adopted especially by the OECD countries (see Table 6.5). These countries accounted for 5.1 mbpd (76%) of the 6.7 mmbpd fall in global demand over the period in question. Of course, oil conservation is likely to prove somewhat more difficult today than in the past since the transportation sector remains overwhelmingly dependent on oil products (gasoline, diesel, kero-jet) which are difficult to substitute. However, over the past few years, there has been a definite pick-up of interest in hybrid motor technology with all the major motor manufacturers declaring their intention of following Toyota and Honda in introducing hybrid vehicles to the market within the next two or three years. These are small beginnings, but investment in alternative technology is rapidly gaining pace, partly owing to higher oil prices. In my personal view, governments need to provide more direct financing of what have hitherto been private sector projects. There is little doubt that current gasoline and diesel engines can be made more efficient, and that hybrid technology could be introduced more rapidly if governments offered appropriate incentives.

Table 6.5
The Global Oil Supply/Demand Balance
(including NGLs) 1979–1983 (mbpd)

	1979	*1980*	*1981*	*1982*	*1983*
Demand					
OECD	42.1	39.0	39.4	37.7	37.0
Old USSR	8.7	8.9	9.1	9.1	9.1
China	1.8	1.8	1.7	1.7	1.7
E. Europe	2.0	1.9	1.0	1.0	0.9
Rest of World	10.9	11.2	9.5	9.8	10.1
Total Demand	65.5	62.8	60.7	59.3	58.8
Annual rise (mb/d)	1.1	-2.7	-2.1	-1.4	-0.5
Supply					
Total non-OPEC of which:	**34.4**	**35.5**	**36.2**	**37.4**	**38.6**
OECD	14.9	15.0	15.1	15.4	16.0
Old USSR	11.7	12.1	12.3	12.3	12.4
China	2.1	2.1	2.0	2.0	2.1
E. Europe	0.4	0.4	0.5	0.5	0.5
Others	5.2	5.8	6.3	7.2	7.6
Total OPEC of which:	**31.5**	**27.6**	**23.6**	**19.9**	**18.6**
Crude	30.6	26.6	22.5	18.8	17.4
NGLs	0.9	1.0	1.1	1.1	1.2
Processing gains	0.9	1.0	0.8	0.9	0.9
Total supply	**66.8**	**64.1**	**60.6**	**58.2**	**58.1**
Implied stock change	1.3	1.3	-0.1	-1.1	-0.7
OPEC market share (%)	**48.1**	**43.9**	**38.9**	**33.6**	**31.6**
OPEC utilization rate (%)	**93.0**	**89.0**	**78.0**	**66.0**	**64.0**

Sources: International Energy Agency (IEA), Varzi Energy estimates.

Looking into the future

One of the key issues for the global economy is the future course of oil prices. While forecasting has always proved difficult, a number of factors are worth mentioning.

First, oil price expectations are definitely rising. In the 1980s to mid 1990s, any price above $25 per barrel was viewed as being too high. Some considered such a price as unsustainable over the medium to long term. Then, as OPEC showed better than expected co-operation and successfully targeted a price band of $22–28 per barrel, market expectations rose gradually to the $25–30 per barrel area. Today, some hedge funds believe that prices could stay above $45 per barrel for some time and even go higher than the current level of around $55 per barrel. Indeed, while not necessarily an accurate forecasting instrument, WTI crude futures stretching out to 2010 currently show prices at over $45 per barrel.

Second, with the world economy and world demand likely to continue on a growth path in 2005 and beyond, there is unlikely to be a sharp drop in the demand for OPEC crude. Non-OPEC supply at best is likely to match rather than exceed annual incremental demand growth, leaving the demand for OPEC crude in the region of 28–30 mbpd.

Third, although OPEC capacity is set to rise, it is unlikely to result in a flood of new oil overnight. This may please many OPEC producers in the short term, but there are definite dangers ahead. If OPEC members fail to re-establish a more comfortable level of spare capacity, the world oil market will continue to suffer from concern about the stability of oil supply. Today, OPEC is virtually producing flat out. Thus an interruption of 2–3 mbpd would send shockwaves throughout the oil market. Such a supply interruption is not entirely inconceivable given the unpredictability of Middle East politics and events in Venezuela, Iraq, Russia and Nigeria over the past year. It would also result in a more negative global economic environment. If, in the event of a major oil price shock, prices were to go back to the real levels prevailing in 1979, a global recession cannot be ruled out altogether.

One key question is why OPEC capacity always seems to rise more slowly than anticipated. Some of the reasons are outlined below:

- *Domestic imperatives:* State companies are constrained by government energy policies, which tend to focus more on defending oil prices and maximizing revenue, than on commercial oil field criteria.
- *Capital constraints:* Whereas private oil companies can decide their own upstream budgets, state oil companies must request funds from their governments, which are often more concerned with using oil revenue to cover current expenditures rather than improving upstream performance.
- *Technology gap:* State oil companies have tended to fall behind in the transfer and application of the latest upstream technology due to limits placed on foreign investment. Certain OPEC member governments cling to an archaic view that, direct involvement on the part of major foreign oil companies in domestic upstream activities reduces their independence of action and infringes their sovereignty and ownership of oil reserves. Such views have definitely and appreciably slowed the rate of upstream investment and technology transfer in those countries, with damaging consequences for their production capabilities.
- *Going for the most easily accessible oil without due regard for longer-term factors:* There has been a tendency to focus on big field developments and to produce from the first discovered pay zone without much attempt to explore lower depths or even, in some cases, to put sufficient emphasis on proper field delineation.
- *Lack of incentives:* State oil company employees are rarely given incentives to perform well and raise productivity.

The general conclusion is that these state oil companies have tended to act rather like the government bureaucracy, and have provided little dynamism to upstream activity.

In my personal view, OPEC members have spent far too much time thinking about prices rather than about the prospects for future world oil supply. OPEC's greatest long-term enemy is the growing concern about supply reliability, which is being voiced globally As a result of the price hike of 1979–81 there was a wave of energy conservation and oil substitution measures by the major oil-importing countries. Such measures could well be implemented again in future years if supply stability again tops the agenda in Western capitals and in countries such as China and India. It is my earnest hope that forthcoming OPEC meetings will address this issue in a coordinated fashion and on an urgent basis.

Finally, I would make a plea for transparency. While OPEC members have become more transparent than in the past about their production numbers, there are no reliable independent estimates of proven and probable reserves. As long as this is the case, certain observers, rightly or wrongly, will continue to insist that OPEC capacity is unlikely to rise much further and that the world is on the verge of a catastrophic global oil shortage. At the very least, greater transparency on upstream data and future capacity expansion would help assure the world that given the right conditions OPEC would be able to raise its capacity to substantially higher levels. It would certainly help politicians and central bankers to make more informed judgments about the potential impact of global oil fundamentals on the prospects for the world economy in the years to come.

Conclusion

In summary, the following conclusions may be drawn from the foregoing discussion:

- Oil and politics have always been intertwined. Oil has played a key role in the world industrialization process. The really big oil reservoirs have been found in a relatively small number of countries, which are concentrated overwhelmingly in the Middle East. Oil not only plays a vital role as a source of fuel for the modern military machine, but also accounts for up to 10% of international trade and therefore has an impact on currencies and other commodities.

- There are several similarities between the price spike of 2004 and previous oil price shocks in 1973 and 1979–81. Global demand has been rising strongly, OPEC capacity is tight and some observers are reviving the issue of long-term reserve replacement.
- However, there are also differences with previous oil price shocks. Above all, the global oil market is much more flexible and better able to weather unexpected developments. The rise of paper markets has resulted in greater transparency. Consuming countries have accumulated a huge volume of strategic stocks, and the global economy uses much less oil per unit of GNP than it did at the end of the 1970s. Moreover, high consumer taxes on petroleum products in many countries mean that rising crude prices have a much smaller impact than in the past.
- Global economic sentiment appears somewhat fragile, given continued concern about global terrorism, rising oil prices and the loss of momentum in key economies, especially in the OECD. However, the world economy will continue to grow in 2005, albeit at a slower pace. Oil prices are still way below the real highs of 1979–81 and the world is currently less dependent on oil as a source of energy than previously.
- Higher oil prices do carry some potential longer-term benefits, such as encouraging higher exploration activity, greater investment in alternate energy sources and better energy conservation.

Overall, it may be concluded that the global economy seems in a much better position to withstand higher oil prices than in past years.

7

The Influence of Markets and Technology on Regional Oil and Gas Reserves

Kjell Aleklett and Colin J. Campbell

The most important point to realize about oil and gas is that these energy resources were formed in the geological past, which means that they are finite and therefore subject to depletion. The formation of oil reserves took place under very special conditions of extreme global warming, with the bulk of the oil in production today coming from just two brief epochs, 90 and 150 million years ago. On the other hand, gas reserves formed under various conditions at different times in the geological past, but being in a gaseous state, much of it escaped from the reservoirs over time. Lastly, it should be pointed out that these two petroleum resources are mobile in nature—once found they flow readily from the wells, often under their own pressure. Therefore, the oil and gas extraction process differs greatly from the laborious mining of coal and minerals where concentration rather than occurrence is the key factor.

In forecasting oil supply, it is important to adopt the mindset and analytical procedures that apply to the specific conditions of the material in question. These conditions differ greatly from those ordained by the precepts of classical economics, which primarily relate to human endeavors free of resource constraints. If wheat prices rise, the farmer will naturally plant more in the next sowing, such that the market price will self-adjust to deliver an optimal value and stimulate appropriate reactive behavior. This explains the initial response by governments to concerns about security of oil supply, which is to press for the liberalization of markets. It is a classic but flawed response, which fails to recognize a curious irony that relates to the extraction of a finite resource, namely: "the better you do the job, the sooner it is exhausted."

All this seems so obvious and yet the subject of depletion is not widely understood, and its importance is not sufficiently recognized, given that the world so heavily depends on a secure supply of oil and gas. The explanation for this failure falls under three broad headings: first, the unreliable nature of public data; second, the pernicious influence of outdated economic theories, which influence political responses; and third, a broad spectrum of vested interests, bent on denial and obfuscation for individual advantage.

It is time to open a new page and shed new light on the subject that must come to the top of the agenda as the first half of the Oil Age draws to a close. It is a large and complex subject, and only the essentials will be summarized in this chapter. For the purposes of general description, it is convenient to treat oil and gas together, save where there is a need to distinguish the two phases.

A view of the planet from outer space would show water covering much of the surface but a closer look would soon distinguish the oceans and seas from the lakes and ponds. In the case of oil, we are dealing not with an ocean or a sea, but rather a lake or pond, having the very special conditions required for its formation. A glance at the oil map confirms that oilfields are generally clustered together in a few preferred places, which are separated one from another by vast barren tracts. They are barren for well-understood geological reasons, which have become progressively refined since the Oil Age dawned 150 years ago when drilling commenced on the shores of the Caspian Sea and in Pennsylvania.[1]

The North Sea Experience

The North Sea is an example of this evolution of knowledge and experience, because information about it is relatively comprehensive and transparent.

The first step towards the North Sea discovery was a well that was drilled by Shell as an exhibit in an industrial fair in Holland in 1937, which to everyone's surprise, encountered oil indications. A second step followed 20 years later, also in Holland where a well was accidentally deepened due to a weekend communications failure and entered continental Permian sandstones, which had not been considered even remotely prospective. Again, to everyone's surprise, the well encountered a major accumulation of gas that was formed by the natural coking of deeply buried coal deposits. This opened up the giant Groningen gas field, which in turn prompted successful exploration in the adjoining North Sea, extending into British waters.

During the 1960s, eyes began to turn northwards in the hope of finding more gas. The first challenge was to establish ownership of the offshore waters between the contiguous countries, principally Norway and Britain, an issue that was duly resolved by a treaty drawing a median line between the coastlines. In those days, seismic technology was relatively primitive so it was not possible to evaluate deep geological environments. The hope was to find more gas, such as that found at Groningen. However, attention shifted to oil, based on minor discoveries in Denmark and the United Kingdom. Norway granted concessions to the southern end of its continental shelf in 1965, and the early seismic surveys showed a cluster of large dome-like structures, capped in Cretaceous chalk, similar to the famous White Cliffs of Dover. It did not appear to be a promising reservoir, but again to everyone's surprise, the second well yielded the giant Ekofisk Field in 1969. Just here, the chalk had especially favorable characteristics, due to slumping at the time of deposition, while the underlying Upper Jurassic strata contained rich source rocks at a depth of prime generation. It was almost considered one of the Wonders of the World.

This discovery stimulated the search for oil in northern waters, and advances in seismic technology supported by massive computing power began to reveal the deep geological environment, identifying a Jurassic rift, which contained both reservoir and source rock. This new oil play opened in the 1970s, with a string of major discoveries straddling the UK–Norway median line. Once a major field had been identified, it became relatively simple to find others.

Another very important technological breakthrough came in the 1980s when advances in geochemistry made it possible not only to identify the source rocks with the help of isotopic signatures but also to model and map the trends of prime generation.

Exploration has continued to the present day and there have been technological advances in virtually all aspects of exploration and production, but once the larger fields were found all that remained were ever smaller ones. Exploration drilling peaked in 1990 with 180 wells drilled but this number has now fallen to around 40, simply because there are fewer viable prospects left, and all the advances in knowledge have made it much easier to determine what is viable and what is not. Looking ahead, it may be concluded that exploration will end within about twenty years. Discovery peaked in 1973, and has declined ever since. It is obvious that production has to mirror discovery after a time lapse. It accordingly peaked in the year 2000 and, barring some anomalous spike, is set to decline at almost 7% a year, as now forecast by both the British and Norwegian governments (see Figure 7.1). The major oil companies perceive this trend very clearly and are selling off their dwindling interests to smaller companies, partly to delay the abandonment charges that are not assignable. With their lower overheads, the smaller companies can still profitably extract the tail-end production, but the amounts added are insignificant in a national context.

Figure 7.1
Past and Estimated Future Discovery and Production of Oil in the North Sea

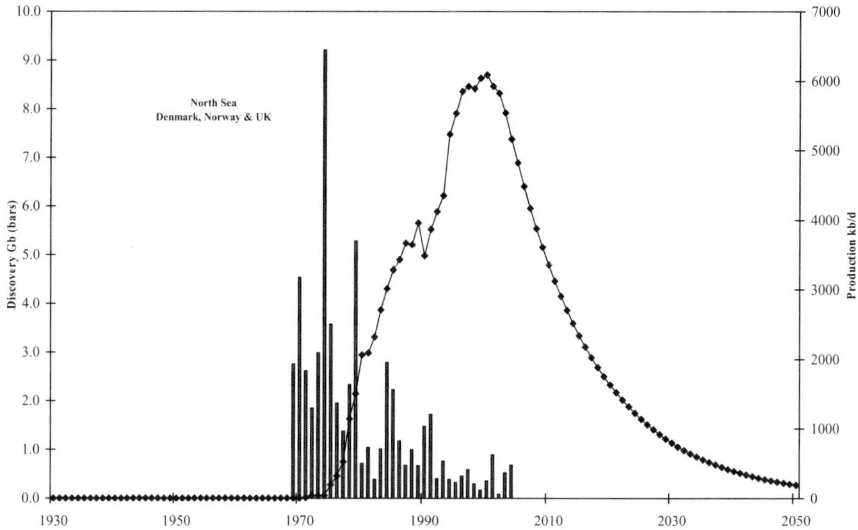

Source: Based on official data from Denmark, Norway, the United Kingdom and the ASPO database.

Those with an economist's mindset often say that the development of the North Sea was a response to the oil shocks of the 1970s, when in fact the critical first giant field was found prior to the shocks. This highlights an important contrast between the mindset of the explorer and the economist. The explorer's task is to find prospects having the right natural geological conditions. At first, he has to work with limited knowledge and try alternative ideas. He is prepared to face failures, and therefore, from his point of view, discovery is a transcendental event. At the outset, the size of the discovery may not be known accurately, but he would say that in terms of discovery, all the oil ever produced from the field, under any technological or economic conditions is attributable to that original borehole. He might add, "Whether you are rich or poor, you have to be born to have a career of any sort."

The economist looks at the same situation with very different eyes. He is interested only in the financial flow. The original discovery may cost relatively little, since most of it is offset against taxable income. The major investments followed later before the golden moment of payout when he could relax. After payout, it was easier to invest more to hold production as high as possible for as long as possible, in order to give a steep terminal decline. The principles of discounted cash flow apply such that money today is worth more than money tomorrow. For the economist, the "discovery" of a single field takes place in many steps, being a continuum measured in money.

The engineer who is responsible for the actual development and production after the discovery, also has his viewpoint. He is by nature rightly cautious, declaring the minimum reserves needed to justify the investment and yield a profit. He works on a step-by-step basis, counting each phase as it is committed. Huge amounts of money are at stake and his reputation suffers if results are below expectations while he basks in glory if they are better.

Finally, there is the manager, who tries to even out the assets of his company. He knows that discovery is episodic, a fact that is not accepted well in the stock market. He also tries to minimize his tax, and sometimes seeks to be less visible. So, he too is motivated to report the minimum and then revise the reserves up over time. This suits the stock exchange rules, which are formulated in order to prevent fraudulent overstatement, but favor under-reporting as commercial prudence. The Securities & Exchange Commission (SEC) recognizes *proved producing reserves,* being the estimated future production of current wells; and *proved undeveloped reserves*, referring to the anticipated production from drilled infill wells even before they are actually drilled.

The Status of World Depletion

This brief description of one productive area, or one of the "ponds" on the global map, and the explanation of the mindsets of the different factions responsible, sets the scene for an outline of the world situation today. The starting point is to determine the discovery trend through the eyes of the explorer. Discovery at any point in time is the sum of past production and the estimated future production of known fields, termed *reserves*. It sounds like a simple matter to resolve but in fact is not easy at all. Five seemingly simple questions must be asked and the available database covers individual fields, not country totals:

1-What was found?
There are many different categories of oil, each with its own costs, characteristics and depletion profile. Some categories are cheap, easy and fast to produce while others are precisely opposite.

2-How much was found?
There are no particular scientific challenges in determining the size of an oilfield early in its life, albeit within a range of quantifiable uncertainty.

3-How much was reported as found?
Here lies the main difficulty in distinguishing the technical estimates from what is expedient to report for financial and political reasons.

4-When was it found?
Here, we have to backdate revisions to the original discovery.

5-Where was it found?
This is easy, once we have answered the other questions.

In the past, the oil companies, which have been diligent in complying with SEC rules and meeting the expectations of the stock market in terms of reported "reserve replacement," were hesitant to address the issue of depletion. However, in 2002, ExxonMobil released a key plot based on properly backdated full-field reserves.[2] It showed that discovery peaked in 1964 and has been in relentless decline ever since. Evidently, the world started finding less than it consumed in 1981. The record of actual discovery as reported by M.G. Salameh is given in Table 7.1. These numbers can be compared with the actual production in Figure 7.2.

Table 7.1
Annual Discovery Excluding USA and Canada

1992	7.80 Gb	1998	7.60 Gb
1993	4.00 Gb	1999	13.00 Gb
1994	6.95 Gb	2000	12.60 Gb
1995	5.62 Gb	2001	8.90 Gb
1996	5.42 Gb	2002	9.00 Gb
1997	5.92 Gb	2003	2.27 Gb

Source: M.G. Salameh, "How Realistic are OPEC's Proved Reserves," *Petroleum Review,* August 2004.

Figure 7.2
The Growing Gap

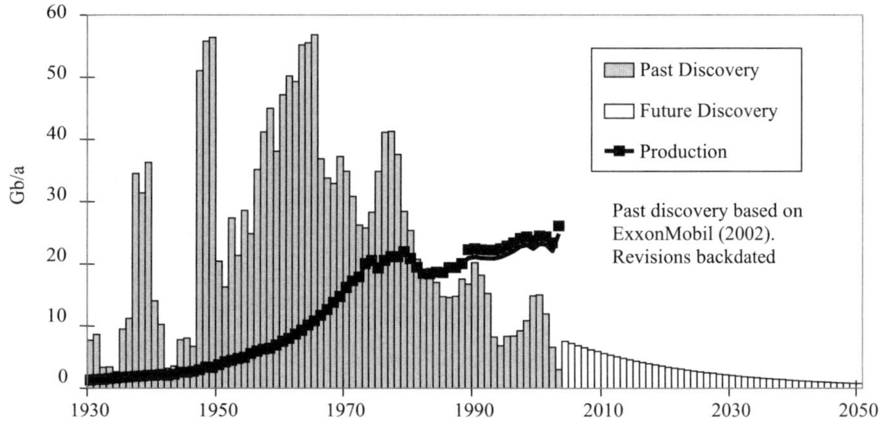

Note: This graph is based upon data from ExxonMobil with production superimposed. Extrapolation of the long downward trend suggests that about 150 Gb is left to be found.

Source: H. Longwell, "The Future of the Oil and Gas Industry: Past Approaches, New Challenges," *World Energy* March 5, 2002.

The discovery table indicates an average of about 7 Gb (billion barrels) a year, with 2003 being an exceptionally low year, when only one barrel was found for every nine consumed. The discovery spikes in 1999 and 2000 related to two large individual discoveries in Kazakhstan and Iran, the latter possibly being the re-announcement of an earlier find. The extrapolation of this long downward trend suggests that about 150 Gb is left to find.

However, before this number is accepted, the questions listed above must be answered to ensure that it is reasonably possible to report sound technical numbers.

It is common practice to distinguish *conventional* from *non-conventional* oil. Unfortunately, there is no standard agreement on the boundary of distinction, adding greatly to the confusion. Therefore, a new term, *regular conventional oil*, may be coined, which by definition excludes the following categories:

- oil from coal and "shale" (actually immature source rock)
- bitumen and derivatives (defined on viscosity)
- Extra-heavy oil (defined on density - $<10^{\circ}$ API)
- heavy oil (10-17.5°API)
- deepwater oil (>500m)
- polar oil
- natural gas liquids from gas plants

Regular conventional oil, as so defined, has contributed most oil to-date and will dominate all supply far into the future. It accordingly controls the critical date of peak production and the onset of decline. The focus will be on this category before considering the other categories.

Appendix 7.1 at the end of this chapter gives an assessment of past production and reserves by country (duly updated for 2004). The starting point is the list of reserves as reported by the *Oil and Gas Journal* (and largely reproduced by BP in its *Statistical Review*). The reported number may be above or below the best estimate of what is actually contained in the reservoir, and has to be multiplied by a factor to respect information contained in various industry databases. The estimates are to be regarded as a point of departure, to be corrected and improved by new information as it becomes available.

The greatest uncertainties relate to the Middle East/Gulf Region, which is defined here to comprise Abu Dhabi, Iran, Iraq, Kuwait, the Neutral

Zone and Saudi Arabia. The starting point in this evaluation is to revert to the reserves reported by the private companies prior to nationalization, assuming that they were conservative estimates complying with SEC rules, being perhaps between 20% and 30% below the best estimates of what was actually there. In 1985, Kuwait added 50% to its reported reserves although nothing particular had changed in the oilfields. This action prompted Venezuela and eventually other countries in the region to retaliate with massive increases to protect their own OPEC quotas, which were partially based on reported reserves. At first, it seemed that these increases were merely making adjustments for the conservative nature of the earlier reports, but the fact that they have barely changed since then, despite substantial production, leads to the conclusion that in fact they represent initial reserves rather than remaining reserves. From the standpoint of an OPEC quota, it may have made much sense to have a relatively fixed number, thus avoiding the need for frequent re-negotiation. Al-Husseini partially confirms this point, reporting that Saudi Arabia has *proved developed reserves* of 130 Gb, which is half the reported number.[3] Without access to the field-by-field details, it is not possible to be certain, and no doubt the present estimates will be subject to correction, but they probably represent a good basis for planning longer term strategies and policies.

Another weak element in the public data is the fact that many countries report unchanged estimates, sometimes for an indefinite number of years. It is implausible for production to exactly match new discovery or reserve revision. Accordingly, in the absence of other information, it is well to reduce the reserves of such countries by the total production of the unchanged period. As many as 68 countries reported unchanged reserves in 2003.

Modeling Depletion

Having assembled at least a working hypothesis of the resource base, it is possible to turn attention to modeling depletion in order to forecast future production. For this purpose, we can divide the countries of the world into three groups:

- Post-midpoint countries are those that have produced more than half their assessed ultimate recovery.
- Pre-midpoint countries are those that have not yet reached midpoint.
- Middle East countries, which have a large proportion of what is left and low depletion rates (annual production as a percentage of the remainder left to be produced from known and new fields).

There are many different scenarios for modeling future production. As a starting point, it seems clear that the Post-midpoint countries, which are past their peak and already firmly into decline, will continue to decline at their current depletion rate.[4] There may, of course, be occasional and unexpected up or down spikes, but on average this seems a reasonable assumption.

Most of the Pre-midpoint countries are now within no more than a few years of midpoint, so the assumptions are not very critical to the longer term. On balance, it seems reasonable to assume that production to midpoint will be flat, based partly on the limits of installed facilities, unless local circumstances suggest otherwise.

The most difficult challenge is to forecast future Middle East production. Previously, the assumption was that the region could, and would, adopt a swing role, making up the difference between world demand under alternative scenarios, and what the other countries can deliver under the model. However, this no longer seems plausible following the invasion of Iraq. Although these countries have relatively low depletion rates, suggesting a physical capacity to increase production, it seems less likely that this will in fact happen for the three following reasons:

- Most of the region's oil is contained in a small number of giant fields found long ago, which are now heading into decline. It follows that it is a major challenge to offset this decline by bringing in the much smaller fields awaiting development, and a much greater challenge to add net capacity.

- The industry in these countries is mainly in state hands and lacks the indirect subsidy available to foreign companies, which are able to offset exploration and operating expense against taxable income.
- These countries have no motive to make the massive investments necessary to increase production, which would serve only to reduce their revenues by lowering oil price, and deplete the reserves more quickly.

Oil revenue is an important element in the budgets of these countries, and their governments have no good reason to squander them, save perhaps to reduce the threat of further invasions. Therefore on balance, it seems likely that regional production will remain rather constant, until the component countries reach their depletion midpoints or until their depletion rates rise to more normal levels.

It has been observed that oil and politics are never far apart. Certainly, many analysts see a link between oil and the invasions of Afghanistan and Iraq.[5] Some fear that both Saudi Arabia and Iran are at risk of invasion for the same reason under whatever pretexts may be developed. The issue of depletion is important in this connection. The EIA (2004), an arm of the US Department of Energy, for example, claims that the Middle East has a productive capacity of 45 million barrels per day (mbpd)—more than double the current production of about 18 mbpd. If this is true, the Middle East might indeed be a valuable prize to be taken by military means—the word "prize" being taken from a speech by US Vice President Dick Cheney.[6] However, it is a very implausible estimate. The saddest of all ironies would be if thousands were to die in abortive military engagements because of the failure to grasp the essential impact of depletion.

If the evaluation of each country under these models is summed up, the peak of global production, as computed, arrives in 2005 for *regular conventional oil* and a few years later for all categories. However, this is no more than a shallow peak, so that relatively small changes in the estimates and modeling could shift it by a few years. Even so, the evidence suggests that, by 2010, the terminal decline will not only have set in, but will also be perceived to have done so.

Natural Gas and Non-Regular Oil

As stressed above, *regular conventional oil* drives the peak, but it is worth commenting briefly on natural gas and the other oil categories.

Natural Gas

Natural Gas was more widely formed in nature than was oil, but that advantage is balanced by the fact that more has escaped over geological time. The greater mobility of gas means that as much as 80% of what is in the reservoir is recoverable, compared with an average of about 35% for oil. The depletion profile is also very different. An uncontrolled well would deplete an accumulation very quickly, so in practice, production is normally set far below capacity to deliver a long plateau rather than a peak. In an open market environment of short term trading, the end of the plateau can come abruptly and without price signals as the United States is now discovering. Gas is more costly to transport than oil, meaning that the market is more regional in character and is based on pipeline distribution systems. As oil prices rise with advancing depletion, attention is now turning to gas liquefaction, allowing stranded gas deposits to be brought into the market, although as much as 11% of the energy is lost in the liquefaction and transport.

In global terms, it seems likely that production will rise to a plateau of 130 trillion cubic feet per annum (Tcf/a) lasting from about 2015 to 2040, followed by a steep decline.

In addition, there are substantial quantities of *non-conventional* gases, such as coal bed methane, which already supply about ten percent of US needs. Production is expected to grow slowly in the years ahead. Much misplaced research has been devoted to hydrates found in polar and deepwater regions, but since the gas occurs in a solid phase, it is unlikely to have accumulated in commercial quantities. It has been rightly said that hydrates are the fuel of the future and are likely to remain so.

Heavy Oils

The resources of various categories of heavy oil are very large, but extraction is slow, costly, and in some cases environmentally damaging. The net energy yield is low or negative, but production is expected to rise gradually in the years ahead, with supply being dominated by production from the world's two largest deposits in Canada and Venezuela. Liquids may be synthesized from coal under the Fischer Tropsch method developed in Germany during the Second World War and still used on a large scale in South Africa. The mining and retorting of oil shales have been attempted in several places but none of the ventures has proved commercially successful so far. That may change in the future as oil prices rise, but the global supply is unlikely to be significant.

Deepwater Oil and Gas

In earlier years, the deepwater domain was considered non-prospective, being too far from land to contain source or reservoir. However, technological progress has brought it within reach, and a few productive areas with exceptional geological environments have been found in the Gulf of Mexico and along the margins of the South Atlantic. Local source rocks developed in early rifts in these places as the continents moved apart, and reservoirs were provided by sands carried down the continental slope in turbidity flows (analogous to submarine "avalanches") which were in turn cleaned of fine-grained material by the winnowing action of long shore currents. A remarkable combination of geological circumstances is responsible for these reserves.

In other areas, especially in the Eastern Hemisphere, deltas locally extend into deep water, but in such cases, source rocks are confined to the delta-fronts themselves and are likely to be lean and gas prone.

Overall, the deepwater domain is expected to have an endowment of about 60 Gb, with production peaking at about 7.5 mbpd around 2014. Gas is barely economical to produce in the deepwater environment.

However, the very fact that companies are willing to explore in these extreme conditions tells us that there are no better alternatives left.

Polar Oil and Gas

Antarctica is not very prospective in geological terms and is in any case closed to exploration by agreement. By contrast, the Arctic regions are more promising with some large sedimentary basins. The evidence to date suggests that they are primarily gas prone due to vertical movements in the geological past under the weight of fluctuating ice caps, which depressed the oil source rocks in the gas-generating zone. Furthermore, these movements served to destroy seal integrity. Although this is the general condition, certain anomalous areas such as around Prudhoe Bay in Alaska and parts of Siberia are oil-bearing, possibly having been buttressed against the vertical movements by local tectonic elements. Accordingly, the Arctic regions have a substantial gas potential, but production is constrained by the high costs associated with the remote and hostile environment.

Natural Gas Liquids

Many oilfields possess associated gas caps, which contain dissolved liquids that condense when brought to the surface, being termed *condensate*. They are commonly metered with oil and for convenience are treated with oil for the purposes of this exercise.

In addition, liquids may be extracted from gas fields by dedicated plants, which may be usefully distinguished insofar as they will deplete in parallel with the gas. Production is currently running at about 2.7 Gb per annum and is expected to reach almost 4 Gb per annum by the gas plateau, forming an important element in the global mix.

Figure 7.3 illustrates the global depletion profile of all categories as currently modeled. In case production has to be maintained at higher levels for longer, especially due to more than anticipated production from the Middle East, the subsequent decline would be that much steeper. It would in fact make a bad global situation worse.

Figure 7.3
Oil and Gas Depletion Profiles
(2004 Base Case)

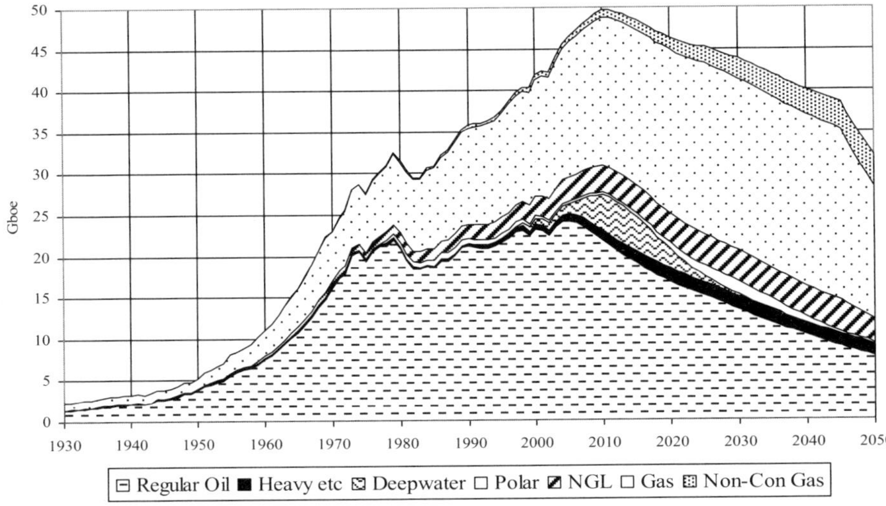

Note: Past and forecast future production of oil and gas, based on the depletion model and estimates as given in the Appendix to this chapter.

Source: Forecast by the authors.

Oil and gas depletion is driven primarily by an immutable resource base, which was set in the geological past, and it is also influenced by the equally immutable physics of the reservoir. Technology and economic factors, which essentially reflect human behavior, can accelerate or slow the rates of depletion within limits, also causing short-term anomalies, but their overall long-term impact is comparatively small. The exploitation of oil and gas has already witnessed great technological progress, such that it is already a highly efficient process, meaning that there is ever reducing scope for still further improvement. The main impact of future economic intervention is by cutting demand, such that the resource might last longer. Making additional investment would have little effect as production is close to its physical capacity. In desperation, more exploratory drilling may be undertaken in areas lacking sound prospects. Some positive surprises may be encountered but the global impact will be negligible.

Potential and Constraints for the Gulf

The Middle East is certainly the critical area for world supply and it is also being subjected to intense political pressures both internally and externally.

The Energy Information Administration (EIA) of the US Department of Energy (DOE) claims in the *International Energy Outlook 2004*, that the Middle East has a productive capacity of 45 mbpd, more than double current production of about 18 mbpd. The Asia Pacific Research Center has recently had a workshop in preparation for the *Energy Outlook 2005* at which the Chinese representative claimed that his country will need to import between 5.1 and 5.5 mbpd on the basis of its current economic growth. If domestic exploration fails to deliver, China will need to increase the import by an additional 2 mbpd. At the same workshop, a report from the University of Tokyo claimed that the Asia Pacific region will need to import 26 mbpd by year 2025. The numbers for Europe and USA were respectively 6 and 7 mbpd.

According to our estimation, the Middle East/Gulf region cannot come close to fulfilling these needs. In February 2004, Saudi Aramco presented its 50-year crude scenarios. A maximum sustainable capacity scenario, with a production of 12 mbpd by 2015, would call for a reserve replacement of 35 billion barrels from 2033 onwards. This is far from the 22.5 mbpd that the EIA claims Saudi Arabia can produce.

The Second Half of the Oil Age and its Impact on Markets

The First Half of the Oil Age has lasted 150 years since the first wells were drilled on the shores of the Caspian and in Pennsylvania. The cheap, convenient and abundant supply of energy from oil facilitated the growth of industry, transport, trade, and agriculture, which made it possible for the world's population to expand six-fold exactly in parallel with oil. In addition, this easy energy led to the growth of large amounts of new financial capital, which in turn gave rise to the subject of Economics by which to manage investment, wealth and finance.

In short, the financial system allowed commercial banks to lend and charge interest on loans that exceeded the amounts placed on deposit, which in turn stimulated new loans. Although intrinsically unstable, the system created its own collateral from the resulting prosperity and economic growth. However, it is well to remember that it is not money but the underlying energy that makes the world go round.

An additional subtle and little understood financial construction was the role of world trading currencies, comprising previously the pound sterling and now the US dollar. Many foreign countries hold and have held national reserves in these currencies, which are widely used for trading transactions—especially in oil. Such transactions are commonly handled by banks based in the home country of the prime trading currency, becoming effectively book transactions which give the banks involved the unseen collateral for issuing still further debt. In fact, the cost of the import of physical oil into the United States has been almost exactly matched by the extension of domestic credit. As oil prices rise, so more and more funds will accumulate to the benefit of the trading banks. It is not surprising that there are political ramifications.

It is evident from the foregoing discussion that the Second Half of the Oil Age will dawn over the next few years. It will be characterized by the decline of oil and all that depends upon it, including the financial structures outlined above.

The transition will likely be a time of great international tension epitomized by the invasions of Afghanistan and Iraq, and the world reactions thereto. Russia is an important element in the complex unfolding situation. Its production collapsed with the fall of the Soviet Union, followed by the brief emergence of new Russian oil companies being run on Western economic practices. Now, it seems that the government is re-exerting its grip on the industry, recognizing that it controls key national assets that need to be preserved. Accordingly, it is likely that production will now level off, reaching a second peak around 2010 before declining gradually.

It seems likely that an increasing number of producing countries will follow the example of the previous Iraqi regime and trade oil in Euros or other local currencies to benefit from some of the hidden tribute now enjoyed by the United States. If so, the strength of the dollar and eventually the world stock markets, which are dominated by US interests, are likely to be further undermined.

In short, it may be said that the decline of oil heralds the "end of economics" as currently understood. Its principles and practices successfully delivered economic growth in parallel with rising oil supply, but new methods and practices are needed to manage the long term decline which characterizes the Second Half of the Oil Age. Furthermore, political structures in so-called democratic countries have mainly aimed at fostering conventional economic practices, with particular emphasis on liberalizing markets. These may also have to change in the face of the new circumstances.

The list of consequences is a long one and cannot be identified other than in the most general terms. Yet, whatever the outcome and whatever the uncertainties, it seems certain that the peak of oil supply represents a discontinuity of historic proportions. In fact it is unprecedented because never before has a critical commodity declined from natural depletion before a better substitute took its place.

The world desperately needs to make sensible plans to face this new and little understood situation. For example, the importing countries could cut imports to match world depletion rate. It would moderate world oil prices, by putting demand in better balance with supply. That would allow the poor countries to afford minimal needs; avoid destabilizing financial flows; and, above all, force the consumers to face reality. It would also serve the Middle East by encouraging those countries to conserve the resources on which they so heavily depend.[7] The first step in such a process is to secure more transparent information on what is the true depletion rate. If valid numbers entered the public domain, the nature and impact of depletion would be entirely self-evident. Although knowledge of the situation might cause panic and over-reaction, including perhaps a proliferation of resource wars, many people would prefer to face the reality of their predicament, so that they could begin to plan and prepare a survival strategy for themselves and their families.

Appendix 7.1: Past Production and Reserves (by Country)

WORLD

Unit: Gb (billion barrels)

REGULAR OIL PRODUCTION (2004)

To 2100

Country		KNOWN FIELDS												NEW FIELDS	ALL FUTURE	TOTAL	Revised 11/21/2005			
		Present		Past								Future	Past & Future				DEPLETION		PEAK	
						Reported Reserves														
		kb/d 2004	Gb/a 2004	Total	5yr Trend	World Oil	O&GJ	Deductions Static	Deductions Other	% Reported							Rate	Mid-Point	Disc	Prod
Saudi Arabia	A	8750	3.19	**100**	2%	259.7	259.4	-39.6	0.0	160%	**162**	263	**12.4**	174.5	**275**	1.8%	2015	1948	2013
Russia	B	8950	3.27	**130**	8%	67.1	60.0	-6.3	-37.0	80%	**75**	205	**14.6**	89.6	**220**	3.5%	1996	1960	1987
US-48	C	3560	1.30	**173**	-4%	2[illegible].9	21.9	0.0	-9.0	90%	**24**	198	**2.3**	26.7	**200**	4.6%	1971	1930	1971
Iran	A	3940	1.44	**57**	1%	130.8	125.8	-20.7	0.0	180%	**70**	127	**12.9**	82.8	**140**	1.7%	2013	1961	1974
Iraq	A	2070	0.76	**29**	-4%	115.0	115.0	-8.7	0.0	185%	**62**	91	**9.2**	71.4	**100**	1.0%	2025	1928	2025
Kuwait	A	2050	0.75	**32**	3%	97.3	99.0	0.0	0.0	180%	**55**	87	**2.7**	57.7	**90**	1.3%	2020	1938	2015
Venezuela	D	1879	0.69	**47**	-5%	52.4	77.2	0.0	-30.0	225%	**34**	82	**5.7**	40.1	**88**	3.2%	1999	1941	1970
Abu Dhabi	A	1955	0.71	**19**	1%	68.6	92.2	-10.5	0.0	230%	**40**	59	**5.5**	45.6	**65**	1.5%	2021	1964	2021
China	B	3494	1.28	**31**	2%	15.4	18.3	-2.5	0.0	75%	**24**	55	**4.6**	28.9	**60**	4.2%	2003	1959	2003
Mexico	D	3410	1.24	**32**	3%	14.8	14.6	0.0	0.0	70%	**21**	53	**2.7**	23.5	**56**	5.0%	2000	1977	2004
Libya	E	1550	0.57	**24**	2%	33.6	39.0	0.0	0.0	190%	**21**	44	**5.5**	26.0	**50**	2.1%	2005	1961	1970
Nigeria	E	2350	0.86	**24**	3%	36.6	35.3	0.0	-5.6	175%	**20**	44	**3.8**	24.0	**48**	3.1%	2004	1967	2004
Kazakhstan	B	986	0.36	**6.7**	9%	-	9.0	-0.7	0.0	30%	**30**	37	**8.3**	38.3	**45**	0.9%	2036	2000	2030
Norway	F	2940	1.07	**19**	-2%	9.9	8.5	0.0	0.0	75%	**11.3**	30	**2.2**	13.5	**32**	7.4%	2002	1979	2001
UK	F	1830	0.67	**21**	-5%	3.9	4.5	0.0	0.0	60%	**7.5**	29	**2.4**	9.9	**31**	6.3%	1997	1974	1999

WORLD

Unit: Gb (billion barrels)

REGULAR OIL PRODUCTION (2004)

To 2100

Country		KNOWN FIELDS											NEW FIELDS	ALL FUTURE	TOTAL	Revised 11/21/2005			
		Present		Past							Future	Past & Future				DEPLETION		PEAK	
						Reported Reserves													
		kb/d 2004	Gb/a 2004	Total	5yr Trend	World Oil	O&GJ	Deductions		% Reported						Rate	Mid -Point	Disc	Prod
Indonesia	G	973	0.36	**21**	-5%	5.3	4.7	-0.4	0.0	60%	**7.8**	28	**1.6**	9.4	**30**	3.6%	1992	1945	1977
Algeria	E	1205	0.41	**13**	10%	15.3	11.8	0.0	0.0	95%	**12.4**	25	**2.6**	15.0	**28**	2.8%	2006	1956	1978
Canada	C	1100	0.40	**20**	0%	4.7	178.9	-0.4	-174.8	3100%	**5.8**	25	**0.7**	6.4	**26**	5.9%	1987	1958	1973
Azerbaijan	B	298	0.11	**8.34**	2%	-	7.0	-0.2	0.0	60%	**11.7**	20	**2.5**	14.2	**23**	0.8%	2014	1871	2009
N.Zone	A	597	0.22	**7.07**	-1%	4.8	5.0	-2.4	0.0	95%	**5.3**	12	**1.7**	6.9	**14**	3.0%	2004	1951	2003
Argentina	D	680	0.26	**8.52**	-2%	2.3	2.7	0.0	0.0	75%	**3.6**	12.1	**0.9**	4.5	**13**	5.5%	1996	1960	1998
Oman	H	767	0.28	**7.57**	-18%	4.8	5.5	-1.3	0.0	110%	**5.0**	12.6	**0.4**	5.4	**13**	4.9%	2001	1962	2001
Egypt	E	712	0.26	**9.21**	-2%	2.2	3.7	0.0	0.0	120%	**3.1**	12.3	**0.7**	3.8	**13**	6.4%	1995	1965	1995
Qatar	H	782	0.29	**7.31**	3%	20.0	15.2	-0.8	-25.0	375%	**4.1**	11.4	**0.1**	4.2	**12**	6.4%	1998	1940	2004
India	G	685	0.25	**6.07**	1%	4.9	5.4	-0.3	0.0	120%	**4.5**	10.6	**0.9**	5.4	**12**	4.4%	2003	1974	2004
Malaysia	G	855	0.31	**5.92**	5%	3.0	3.0	-0.9	0.0	75%	**4.0**	9.9	**0.6**	4.6	**11**	6.4%	2002	1973	2004
Australia	G	430	0.16	**6.12**	-8%	3.6	1.5	0.0	-1.0	80%	**1.9**	8.0	**2.0**	3.9	**10**	3.9%	1999	1967	2000
Colombia	D	530	0.20	**5.94**	-5%	1.5	1.5	-0.4	0.0	50%	**3.1**	9.0	**1.0**	4.1	**10**	4.8%	1999	1992	1999
Angola	E	480	0.18	**4.96**	-7%	9.0	5.4	-2.0	-9.9	140%	**3.9**	8.8	**0.7**	4.5	**10**	3.7%	2004	1971	1998
Ecuador	D	518	0.19	**3.63**	6%	5.5	4.6	-0.3	0.0	110%	**4.2**	7.8	**0.2**	4.4	**8.0**	4.1%	2006	1969	2004
Romania	B	102	0.04	**5.83**	-3%	0.5	1.0	-0.1	0.0	110%	**0.9**	6.7	**0.3**	1.2	**7.0**	3.1%	1970	1857	1976
Syria	H	504	0.18	**4.17**	-1%	2.3	2.5	-2.2	0.0	100%	**2.5**	6.7	**0.3**	2.8	**7.0**	6.1%	2000	1966	1995

WORLD

Unit: Gb (billion barrels)

REGULAR OIL PRODUCTION (2004)

To 2100

Country		KNOWN FIELDS											NEW FIELDS	ALL FUTURE	TOTAL	Revised 11/21/2005			
		Present		Past							Future	Past & Future				DEPLETION		PEAK	
						Reported Reserves													
		kb/d 2004	Gb/a 2004	Total	5yr Trend	World Oil	O&GJ	Deductions		% Reported						Rate	Mid -Point	Disc	Prod
Brazil	D	257	0.09	**4.88**	-6%	11.2	8.5	0.0	-12.0	425%	**2.0**	6.9	**0.1**	2.1	**7.0**	4.2%	1995	1975	1986
Dubai	H	350	0.13	**3.99**	5%	1.3	4.0	-2.2	0.0	300%	**1.3**	5.3	**0.7**	2.0	**6.0**	6.0%	1995	1970	1991
Turkmenistan	E	216	0.08	**3.10**	10%	-	0.5	-0.3	0.0	50%	**1.1**	4.19	**1.3**	2.4	**5.5**	3.2%	1998	1964	1973
Trinidad	D	130	0.05	**3.30**	2%	0.8	1.0	-0.1	0.0	85%	**1.2**	4.46	**0.3**	1.5	**4.8**	3.2%	1985	1959	1978
Gabon	E	235	0.09	**3.02**	-6%	2.2	2.5	-0.7	0.0	170%	**1.5**	4.49	**0.0**	1.5	**4.5**	5.5%	1997	1985	1996
Brunei	C	190	0.07	**3.14**	1%	1.1	1.4	-0.9	0.0	110%	**1.2**	4.36	**0.1**	1.4	**4.5**	4.8%	1989	1929	1978
Ukraine	B	80	0.03	**2.72**	2%	-	0.4	-0.1	0.0	40%	**1.0**	3.71	**0.3**	1.3	**4.0**	2.2%	1984	1962	1970
Vietnam	G	340	0.12	**1.14**	2%	1.4	0.6	-0.8	0.0	30%	**2.0**	3.14	**0.4**	2.4	**3.5**	5.0%	2009	1975	2005
Yemen	H	350	0.13	**1.87**	0%	3.0	4.0	-1.5	0.0	340%	**1.2**	3.05	**0.5**	1.6	**3.5**	7.3%	2003	1978	1999
Denmark	F	393	0.14	**1.61**	5%	1.3	1.3	0.0	0.0	120%	**1.3**	2.93	**0.6**	1.9	**3.5**	7.1%	2005	1971	2004
Peru	D	81	0.03	**2.39**	-3%	1.0	1.0	0.0	-0.6	110%	**0.9**	3.25	**0.2**	1.1	**3.5**	2.7%	1988	1861	1983
Uzbekistan	B	134	0.05	**1.16**	-3%	-	0.6	-0.3	0.0	50%	**1.2**	2.34	**0.4**	1.6	**2.8**	3.0%	2008	1992	1998
Congo	E	240	0.09	**1.69**	-2%	1.8	1.5	-0.9	-0.5	210%	**0.7**	2.41	**0.3**	1.1	**2.8**	7.7%	2000	1984	2001
Germany	F	69	0.03	**1.98**	2%	0.2	0.4	0.0	0.0	120%	**0.3**	2.31	**0.2**	0.5	**2.5**	4.6%	1977	1952	1966
Sudan	E	287	0.10	**0.44**	11%	6.4	0.6	-0.3	0.0	50%	**1.1**	1.56	**0.9**	2.1	**2.5**	4.8%	2009	1980	2005
Chad	E	247	0.09	**0.13**	-	-	-	0.0	0.0	-	**1.2**	1.33	**0.7**	1.9	**2.0**	4.6%	2014	1977	2008
Tunisia	E	70	0.03	**1.25**	-2%	0.7	0.3	-0.2	0.0	75%	**0.4**	1.66	**0.3**	0.8	**2.0**	3.3%	1998	1971	1981

WORLD

Unit: Gb (billion barrels)

REGULAR OIL PRODUCTION (2004)

To 2100

Country		KNOWN FIELDS											NEW FIELDS	ALL FUTURE	TOTAL	Revised 11/21/2005			
		Present		Past							Future	Past & Future				DEPLETION		PEAK	
						Reported Reserves													
		kb/d 2004	Gb/a 2004	Total	5yr Trend	World Oil	O&GJ	Deductions		% Reported						Rate	Mid -Point	Disc	Prod
Italy	F	115	0.04	**0.96**	5%	0.7	0.6	-0.1	-0.3	80%	**0.8**	1.74	**0.3**	1.0	**2.0**	3.9%	2005	1981	2004
Thailand	G	154	0.06	**0.54**	8%	0.5	0.6	-0.1	0.0	80%	**0.7**	1.27	**0.3**	1.1	**1.6**	5.0%	2008	1981	2005
Cameroon	E	70	0.02	**1.08**	-5%	-	0.4	-0.7	0.0	110%	**0.4**	1.44	**0.1**	0.4	**1.5**	5.1%	1994	1977	1986
Bolivia	D	35	0.01	**0.45**	5%	0.5	0.4	0.0	0.0	80%	**0.6**	1.00	**0.3**	0.8	**1.3**	1.6%	2016	1966	2010
Bahrain	H	34	0.01	**1.00**	2%	-	0.1	0.0	0.0	60%	**0.2**	1.21	**0.0**	0.2	**1.3**	5.0%	1977	1932	1970
Turkey	H	42	0.02	**0.86**	-5%	0.3	0.3	0.0	0.0	150%	**0.2**	1.06	**0.1**	0.3	**1.2**	4.3%	1992	1969	1991
Netherlands	F	44	0.02	**0.87**	-2%	0.2	0.1	0.0	0.0	40%	**0.3**	1.14	**0.1**	0.3	**1.2**	4.7%	1991	1980	1987
Croatia	B	19	0.01	**0.51**	-3%	0.1	0.1	0.0	0.0	24%	**0.3**	0.82	**0.2**	0.5	**1.0**	1.4%	2003	1950	1988
Hungary	B	22	0.01	**0.69**	-5%	0.2	0.1	0.0	0.0	70%	**0.1**	0.84	**0.2**	0.3	**1.0**	2.6%	1987	1964	1987
France	F	23	0.01	**0.74**	-4%	0.2	0.1	0.0	0.0	95%	**0.2**	0.90	**0.1**	0.2	**1.0**	3.9%	1987	1958	1988
Pakistan	G	62	0.02	**0.50**	15%	0.3	0.3	0.0	0.0	100%	**0.3**	0.79	**0.4**	0.4	**0.9**	5.4%	2001	1983	1992
Austria	F	18	0.01	**0.79**	-1%	0.1	0.1	0.0	0.0	70%	**0.1**	0.88	**0.0**	0.1	**0.9**	5.6%	1970	1947	1955
Papua	G	46	0.02	**0.38**	-8%	0.2	0.2	0.0	0.0	70%	**0.3**	0.72	**0.1**	0.5	**0.9**	3.4%	2007	1987	1993
Sharjah	H	48	0.02	**0.50**	-1%	-	1.5	0.0	0.0	1000%	**0.2**	0.65	**0.1**	0.3	**0.8**	5.6%	1998	1980	1998
Albania	B	6	0.00	**0.54**	1%	0.2	0.2	0.0	0.0	85%	**0.2**	0.73	**0.1**	0.3	**0.8**	0.9%	1986	1928	1983
Chile	D	10	0.00	**0.43**	9%	0.1	0.2	0.0	0.0	400%	**0.0**	0.47	**0.0**	0.1	**0.5**	4.8%	1979	1960	1982

WORLD

Unit: Gb (billion barrels)

REGULAR OIL PRODUCTION (2004)

To 2100

Country		KNOWN FIELDS											NEW FIELDS	ALL FUTURE	TOTAL	Revised 11/21/2005			
		Present		Past							Future	Past & Future				DEPLETION		PEAK	
						Reported Reserves													
		kb/d 2004	Gb/a 2004	Total	5yr Trend	World Oil	O&GJ	Deductions		% Reported						Rate	Mid -Point	Disc	Prod
ME Gulf	A	19362	7.07	**245**	1%	676	696	-82	0	177%	**395**	640	**44**	439	**684**	1.6%	2017	1948	1974
Eurasia	B	14308	5.22	**191**	6%	33	97	-11	-37	67%	**146**	337	**33**	179	**370**	2.8%	2003	1964	1987
N.America	C	4660	1.70	**193**	-2%	27	201	0	-184	667%	**30**	223	**3**	33	**226**	4.9%	1973	1930	1972
L.America	D	7530	2.75	**110**	-1%	90	112	-1	-43	158%	**71**	181	**11**	82	**192**	3.3%	1999	1977	1998
Africa	E	7438	2.71	**83**	2%	109	100	-5	-16	154%	**65**	148	**16**	81	**164**	3.0%	2004	1961	2004
Europe	F	5431	1.98	**47**	-3%	17	16	0	0	72%	**22**	68	**6**	28	**74**	6.7%	2000	1974	2000
East	G	3735	1.36	**44**	-1%	90	18	-3	-1	77%	**23**	67	**6**	29	**73**	4.5%	1999	1967	2000
ME. Other	H	2877	1.05	**27**	0%	32	33	-8	-25	227%	**15**	42	**2**	17	**44**	5.8%	1999	1965	1998
Other	I	647	0.24	**3.8**	7%	-	0	0	0	93%	**1**	5	**5**	6	**10**	3.7%	2009	1956	2007
Unforeseen	J							0			**10**		**4**	14	**14**				
Non-ME		46626	16.9	**682**	1%	378	577	-29	-305	151%	**382**	1063	**103**	484	**1166**	3.4%	1997	1956	2004
WORLD		65988	24.1	**944**	1%	1054	1273	-111	-305	164%	**776**	1720	**130**	906	**1850**	2.6%	2003	1964	2004

Note: Data has been updated to 2004.

Source: ASPO database.

Section 4

Gulf Energy: Growth and Investment

8

The Future of Gas Transportation in the Middle East Region: LNG, GTL and Pipelines

James T. Jensen

Until recently, the high costs of gas transportation had all but excluded the Middle East from cross-border gas trade. In 1993, only 6.5 percent of Middle East gas production crossed an international border, and since 3.7 percent of this represented trade among the constituent members of the United Arab Emirates federation, only 2.7 percent was truly international—the export from Abu Dhabi to Japan of 3.4 billion cubic meters (bcm) of liquefied natural gas (LNG).

However, that situation is changing rapidly. A decade later, in 2003, trade accounted for 15.2 percent of Middle East production and LNG exports had increased tenfold. The International Energy Agency now expects that the Middle East will be the fastest growing source of gas for inter-regional markets over the next two decades.

Factors Affecting the Outlook for Middle East Gas Exports

Three factors may be identified as being primarily responsible for boosting the outlook for gas exports: first, gas-fired combined cycle gas turbine (CCGT) technology has made gas the fuel of choice for power generation throughout the world; second, the rapid growth in gas demand has stimulated interest among previously gas-poor countries and taxed the availability of traditional North American and European gas supplies; and third, technological improvements in LNG and pipelining as well as gas-to-liquids (GTL) conversion have helped to bring down the costs significantly.

CCGT units have revolutionized the use of gas in power generation. The thermal efficiency of the conventional steam boiler is limited to about 38 percent. However, by adding a high-temperature gas turbine on the front end of the steam cycle, CCGT unit efficiencies approach 60 percent. CCGT units also have low capital cost, can be built with short lead times and come in "bite-sized" modules that are well-adapted to market growth. Moreover, in an environment of growing concern over carbon dioxide emissions, the CCGT's combination of high thermal efficiency with a high hydrogen-to-carbon ratio fuel is an added advantage.

The rapid growth in gas use for power generation has created demand in previously gas-poor countries, such as Spain, Turkey, India and China. In addition, it has stimulated gas demand growth in North America and Europe to the point where demand is taxing traditional sources of supply. This is especially true for the United States and the United Kingdom. Demand continues to grow even as indigenous production in the United States and Canada has apparently leveled out and the United Kingdom is facing a decline in North Sea production.

Simultaneously, there have been cost improvements in LNG, gas pipelining and GTL technology, making it possible to move natural gas over longer distances. The relatively lower cost of gas transportation coupled with increased interest in cross-border trade has positioned the Middle East as a major player in the international natural gas trade.

The Outlook for the Middle East Supply and Trade

After seventeen years of relatively minor participation in international LNG trade, Middle East exports increased dramatically in 1994 with the first shipments from a new LNG train built by Abu Dhabi Gas Liquefaction Limited (ADGAS) in the United Arab Emirates and followed shortly thereafter by the startup of the Qatargas 1 project in Qatar. Since 1994, Middle East LNG exports have grown at a rate exceeding twenty-five percent per year and the current exporters in this region include Oman, Abu Dhabi and Qatar. Both Iran and Yemen are pursuing LNG projects of their own. This rapid spurt in Middle East exports is shown in Figure 8.1.[1]

Figure 8.1
Rapid Middle East LNG Expansion and Shift in Export Focus from Pacific Basin to Atlantic Basin (BCM)

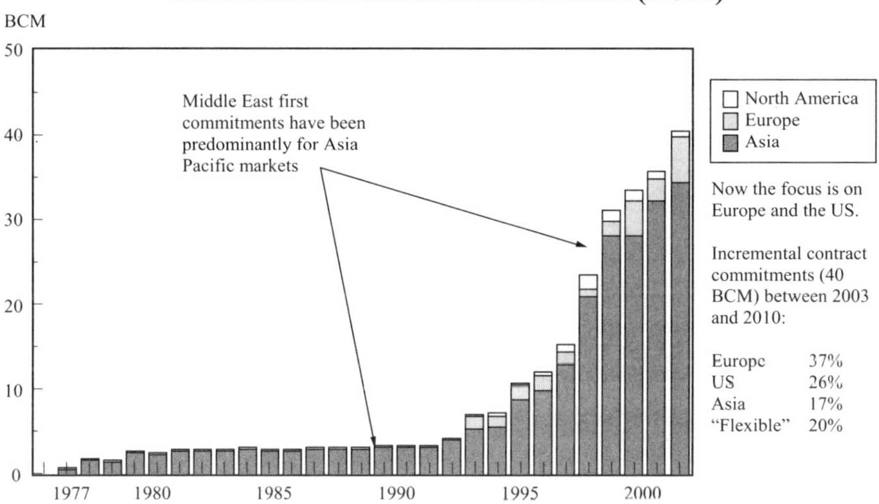

Source: Data from Cedigaz, *Natural Gas in the World* and BP's *Statistical Review of World Energy* various issues.

The initial exports were to the traditional LNG markets in Northeast Asia—Japan, Korea and Taiwan. However, the emergence of European and North American interest in LNG imports has shifted the Middle East's export interest to Atlantic Basin markets. While only a small percentage of 2004 shipments were delivered to Atlantic Basin markets, sixty-three percent of the new contract commitments at the end of 2003 were for Europe and the United States not including another twenty percent with "flexible" destinations—which have traditionally been delivered west of the Suez as well.

The International Energy Agency distinguishes what it terms "major net inter-regional trade" from the common but shorter-haul, intra-regional movements, such as between Canada and the United States or The Netherlands and Germany. By this definition, total "inter-regional trade" in 2002 amounted to 283 bcm, or roughly half of total cross-border trade. LNG, which was essentially all inter-regional, accounted for fifty-three percent of inter-regional trade—amounting to twenty-six percent of total cross-border trade.

Between 2002 and 2030, the IEA expects major inter-regional gas trade to grow three and one half times (see Figure 8.2).[2] In 2002, the Former Soviet Union (FSU) accounted for nearly forty percent of total inter-regional trade via its major pipeline delivery system to Europe. In contrast, the LNG shipments from the Middle East accounted for only twelve percent. However, the IEA foresees a major reversal of the relative roles of the two regions. It expects that the Middle East will account for thirty-eight percent of the growth in trade between 2002 and 2030, while the FSU's maturing trade patterns will account for only seventeen percent of the increment. African growth, both by pipeline and as LNG, will also be larger than that of the FSU.

Figure 8.2 also shows the trade destinations. Europe will remain the largest importer, followed by North America. Northeast Asia, which has been second only to Europe in 2002, is not expected to increase substantially. In the IEA's view, China and India, although they represent interesting new markets, are comparatively small compared with the Atlantic Basin markets.

The emerging GTL technology had its beginnings in South Africa and in Malaysia. Sasol, which has been operating a coal-to-liquids plant for many years in South Africa, pioneered its GTL process at its Mossgas facility in that country. However, Shell built its GTL demonstration plant at Bintulu in Malaysia near its LNG export operations.

Despite GTL proposals for Algeria, Australia, Nigeria and Iran, the Middle East appears likely to become the major supplier of GTL to world markets. Qatar has chosen to be a major source of GTL and has six proposed projects on the drawing boards. While three of these are on hold because of the limited capacity of the design construction firms to take on new projects,[3] all appear likely to proceed at some point in the future. Existing GTL capacity in the two plants amounts to thirty seven thousand barrels per day. Planned projects will add capacity of another nine hundred and thirty five thousand barrels per day. Qatar alone, taking into account the second phase expansion of proposed projects, would account for eighty-six percent of the proposed expansions.[4]

Figure 8.2
Estimated Growth of Inter-regional Gas Trade (Pipeline and LNG) between 2002 and 2030 (BCM)

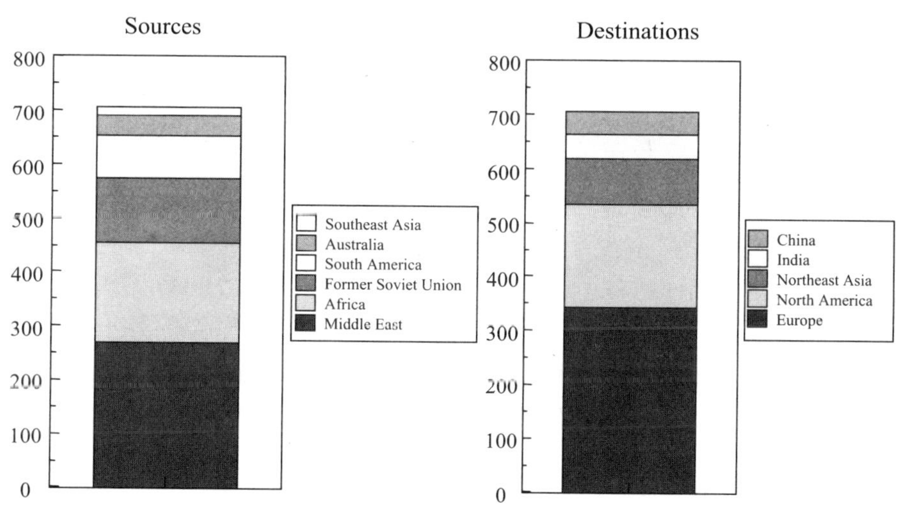

Source: International Energy Agency, *World Energy Outlook 2004*, Figure 4.9, 143.

Long distance pipelining is expensive and the Middle East has been slow to develop pipeline outlets. Some regional projects, such as Dolphin from Qatar to Dubai, exist or are going forward, but a number of inter-regional pipelines are in the discussion stage. Iran now exports to Turkey by pipeline and is in discussions for a pipeline link to the Indian subcontinent. More ambitious proposals to link Iran with Europe are also under consideration.

There is ample supply to support gas export projects in the Middle East. Between them, the Middle East and the FSU account for nearly three-quarters of the world's proved reserves of natural gas—forty-one percent in the Middle East and thirty-two percent in the FSU.

The two largest holders of proven gas reserves in the Middle East are Iran and Qatar (see Table 8.1). Qatar is now a major exporter and Iran aspires to be one. Qatar's reserves are largely contained in its North Field, which together with Iran's South Pars, is the world's largest gas field. Iran's share of the combined field is smaller than Qatar's share, but it has a number of other large non-associated gas fields, as well. However, a significant portion of Iran's proven reserves are in the form of dissolved gas and gas that is located in the large gas caps of producing oil fields. These reserves are more difficult to utilize for export without jeopardizing oil production and recovery. In fact, much of the early South Pars development is designated for re-injection into the oil fields for enhanced oil recovery. Nevertheless, Iran has a substantial potential for export supply.

Table 8.1
Natural Gas: Proven Reserves (year-end 2004)

	Total (in trillion cubic feet, tcf)	*Approximate Percent Non-Associated*
Iran	971	63%
Qatar	910	99%
Saudi Arabia	238	43%
UAE	214	30%
Iraq	112	11%
Kuwait	56	3%
Oman	35	75%
Yemen	17	56%
Other Middle East	18	71%
Total	**2,571**	**68%**

Sources: Reserves: BP *Statistical Review of World Energy*, June 2005; Non-Associated Gas: Author's estimates.

The only other current inter-regional exporters are the UAE and Oman, although Yemen has an LNG export project in the planning stage. Unlike Qatar and Iran, both Oman and Yemen have limited proven reserves that will limit their potential to increase exports over current plans without additional discoveries.

Abu Dhabi was the first LNG exporter from the Middle East and it launched an expansion during the 1990s. However, a large portion of its reserves are in dissolved gas and gas caps which will influence its decision to further expand its exports.

Saudi Arabia has the third largest reserves in the Middle East, but a large portion of them represent dissolved gas. In the early 1980s, Saudi Arabia commercialized much of its associated gas production with the Master Gas System for supplying fuel and feedstock to the Kingdom. In recent years it has been drilling for gas in its Khuff formations, increasing the share of non-associated gas in its reserve inventory. In the early 2000s, it began negotiations with a group of international oil companies on its "Natural Gas Initiative" to further develop its non-associated gas supplies.

Despite its large proven reserve base and its increasing emphasis on gas exploration, Saudi Arabia has shown little inclination to enter gas export markets. This reluctance has been publicly stated by Ali Naimi, the Saudi Minister of Petroleum and Mineral Resources, as an emphasis on keeping the gas within the country for industrial development.[5]

Competition between LNG, Pipelines and GTL

The different economic characteristics of LNG, gas pipelines and GTL exercise the greatest influence in areas where they are the most competitive. The three methods of transportation are especially sensitive to the distance over which the gas must travel and the scale of the operation.

Pipeline gas, once treated to meet quality specifications, requires no further processing and the costs are a direct function of distance. However, the costs may be high, and they are very sensitive to economies of scale.

Large diameter pipelines, which require large markets, can minimize transportation costs. Marine pipelining is more expensive than onshore pipelining, but technological improvements in offshore pipelines enable them to compete in markets that were unthinkable twenty years ago.

LNG requires costly processing, both in liquefaction and in re-gasification at the receiving terminal. Since these costs are incurred regardless of the distance traveled, LNG is at a cost disadvantage for shorter hauls. However, the cost of moving gas in an LNG tanker per kilometer are much lower than moving it through pipelines, and therefore LNG is more competitive for long distance transportation.

GTL technology is difficult to compare with either pipelines or LNG, since it is not only a method of transporting gas, but also a conversion process to a liquid fuel. GTL processing provides high yields of high quality naphthas and diesel fuels, which compete in a very different market from the gas delivered by pipeline or LNG. The quality of the diesel product will more than meet the tighter quality specifications for diesel fuel that are emerging in many markets. While it can be used directly, it may have a higher value as a blending stock for conventional crude oil-based distillates. GTL has higher processing costs and thermal conversion losses than LNG, but its market premium can provide an offset to these costs. It also can be transported in clean conventional tankers at considerably lesser cost than that of moving LNG in the more expensive LNG tankers. Figure 8.3 illustrates the comparative costs of delivering energy in the three forms as a function of distance. It assumes a field with a 9 trillion cubic feet (tcf) reserve capable of supporting production of 900 million cubic feet per day (mmcfd) It assumes gas delivered into the plant or pipeline at a transfer price of $0.50 per million British Thermal Units (mmbtu) and that the yield of higher-quality GTL distillate product can command a 20% premium over crude oil at market (which in the figure is credited against the value of the feedstock).

The 900 mmcfd would supply a 36 inch, low-pressure pipeline, a system representative of the pipelines that built much of the original US pipeline grid. It would also supply a six million ton, single train LNG project. While this would be slightly larger than the four million ton trains that are becoming common, it is smaller than several of the new trains that have been proposed for Qatar. Alternatively, the 900 mmcfd would supply a GTL plant with a capacity of 100,000 barrels per day (bpd).

The estimates in Figure 8.3 are those of the author, using his own proprietary economic models. The original data for these models came from literature sources, but they have been maintained over the years by literature screening and periodic updating discussions with industry sources. The hurdle rates for the individual models vary in an effort to reflect the rate of return that investors might require for that particular investment reflecting its relative risk.

The economic illustrations in this paper commonly use a "cost-of-service" approach for the comparison of costs. This assumes that the delivered price for the energy is built up from the wellhead from costs and utilizes a common assumed into-plant or into-pipeline "cost" of $0.50 per million British Thermal Units (mmbtu).

Actual projects are done on a very different basis – a "netback" basis – which takes a price in the marketplace and deducts transportation and processing costs from the market price to "net back" a transfer price into the plant or pipeline. The way in which the revenue from this netback is divided between the international investors and the host country is defined by the partnership agreement and the tax regime. In the Middle East the national oil company commonly is a part of the agreement and thus may play a dual role in both operations and tax collection. Therefore, the host government can influence the development of the project either through its role as a partner or in modifying the project-specific tax regime.

In comparing the costs of transportation in this chapter, it is important to note that the past several years have seen substantial inflationary pressures operating to increase costs for all three methods of gas transport. Thus the economic comparisons – believed to be valid for early 2005 – can change as a result of these inflationary pressures.

Figure 8.3
Illustrative Costs of Delivering Energy from 900 MMcfd Supply as LNG, High Pressure and Low Pressure Pipeline Gas and GTL ($/MMBTU*)

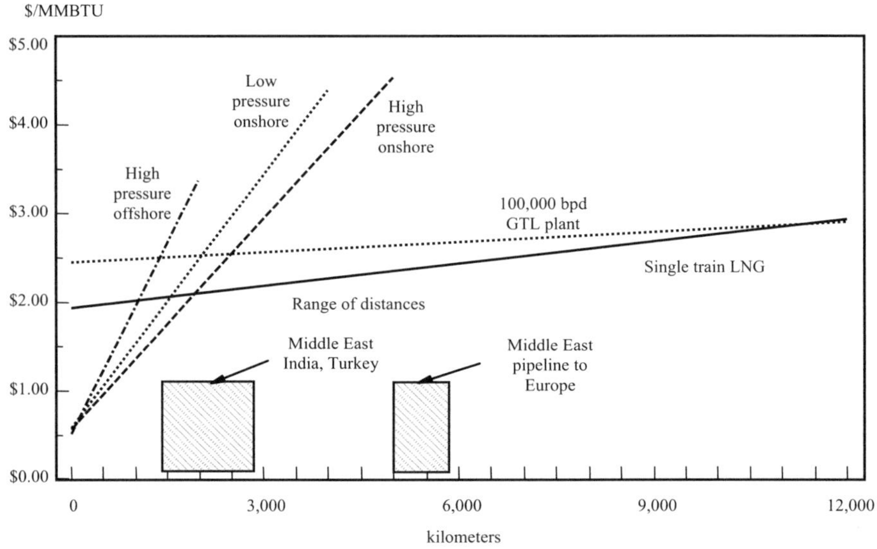

Note: *Assumes $0.50 gas into plant and a quality premium for GTL of $6/BBL credited to cost of gas.

Source: Author's estimates.

As Figure 8.3 indicates, pipelines demonstrate superior economics up to about 1,500 kilometers. In the range of 1,500 to 2,500 kilometers, LNG provides competition for pipelines. Figure 8.3 illustrates the improvement in economics of moving the 900 mmcfd in a high-pressure 28 inch line compared to the 36 inch low pressure line. The low pressure lines operate in the range of 800 to 1,000 pounds per square inch (psi) while the newer high pressure lines may operate at 2,000 psi or even higher. The illustration uses high pressure to move the same amount of gas more efficiently than the low pressure line could accomplish. But high pressure lines demonstrate their real economic impact by moving much larger volumes in large diameter lines. For very long hauls, 42 inch lines improve the economics substantially and even larger diameter lines with higher throughput are on the drawing boards. For longer distances, LNG's only challenger is GTL.

The three distance ranges roughly define where the three systems compete. Local and regional markets are best served by pipelines. The proposed new Dolphin line from Qatar to Abu Dhabi and Dubai is an example of the type of movement that is ideally served by pipeline.

In OECD markets there is some trucking of LNG from receipt terminals to satellite markets, but this is a feature of a well-developed pipeline and distribution infrastructure and is probably not applicable to the Middle East. Similarly, a new technology – compressed natural gas – may also compete in this short haul distance range. However, it appears to be best suited for movements from small offshore fields to nearby large developed markets, and is also likely to have limited relevance for Middle East applications.

As shipping distances lengthen, LNG emerges as a potential competitor to pipelines. Markets such as the Indian subcontinent and Turkey are within a distance range where pipeline and LNG costs may be comparable. Although pipelining does not appear to be competitive with LNG for the more distant markets in Europe as indicated in Figure 8.3, the case shown represents a comparatively small volume movement. Larger shipments via large diameter pipelines bring down transportation costs and may make pipelines competitive in southern and central Europe. In all cases, however, political concerns play an extremely important role in the selection of the transport system, often overriding project economics.

Competition between pipelines and LNG is not new. Algeria pioneered the Atlantic Basin LNG trade in 1964 with its Camel project delivering LNG to France and the United Kingdom. It followed up with the construction of its Skikda and Arzew liquefaction facilities in the mid-1970s and added Spain as a customer of Skikda in 1975. At the time that contract was written, deep-water pipeline technology for a Mediterranean crossing did not exist and LNG was the only alternative. However, in

1977, when the deep water technological problem was solved with the Trans-Mediterranean Pipeline to Italy, all subsequent expansions to Italy and the Iberian Peninsula have been carried out by pipeline.

It was clear at the time the Trans-Mediterranean system was laid that the cost advantage favored the pipeline option over LNG for the shorter distance to Italy. However, both LNG and pipelines have experienced cost changes over time as technology has improved.

The cost improvements in pipelining have come primarily from changes in design, such as high pressures and marine pipe-laying. However, the underlying tendency for material and labor costs to rise seems to be a partial offset to the design improvements. This has been particularly true over the past year as steel prices have risen substantially, providing significant inflationary pressures on pipeline costs.

The perception that the Trans-Mediterranean pipelining was lower cost than LNG probably held true until quite recently. Yet, by 2005, the superiority of the pipelining option is not as clear cut. Figure 8.4 provides an illustration of the comparative economics of pipelines and LNG from Algeria to Italy both from the 1997 and 2005 perspectives. It shows the pipeline as the cheaper alternative in 1997 but indicates that LNG may have won out by 2005. In the illustration, LNG costs have benefited from the cost reduction trends of the last few years, but the use of the same pipeline design in the two periods would have resulted in a cost increase (assumed to be about 25%) due to higher material and labor costs. While Figure 8.4 reflects inflationary pressures on pipeline costs, it does not factor in similar inflationary pressures on LNG, which may be imminent.[6] Should these inflationary pressures develop, it would serve to restore some of the earlier competitive advantage to cross Mediterranean pipelining.

It is difficult to make direct comparisons of the economics of the two gas delivery systems. The pipeline in the illustration is designed to deliver to Rome (although the actual Trans-Med system goes farther north) and would be delivering gas along the way in Sicily and southern Italy. The

LNG project comparison is limited to delivery of re-gasified LNG at the La Spezia terminal near Genoa, without any ongoing further distribution. Thus their destinations are not strictly comparable. Also, the Trans-Mediterranean system underwent substantial expansion in 1996 providing some potential for economies of scale over the original design.

Figure 8.4
Illustrative Costs of Delivering Gas from Algeria (Hassi R'Mel) to Italy

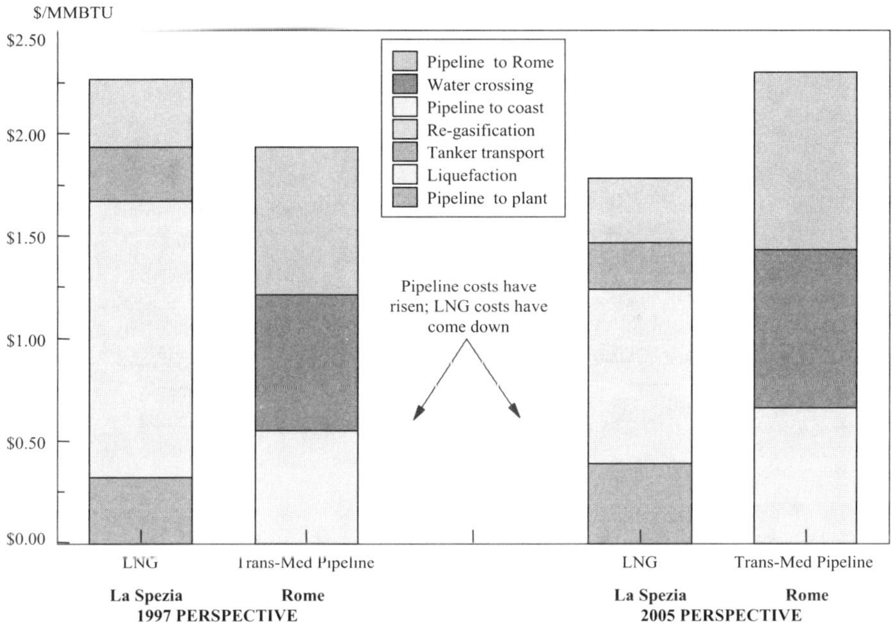

Note: Estimates assume original pipeline design sizing, two train LNG plants, present construction costs and 90% pipeline load factor operation.

Source: Author's estimates.

The comparative economics of a Mediterranean crossing to Spain more clearly favor pipelining. The construction of the Mahgreb Pipeline to Spain was also completed in 1996. Figure 8.5 compares the economics of the two systems. Despite the increase in pipeline costs and the reduction in LNG costs, the pipeline alternative is favored in both cases.

Figure 8.5
Illustrative Costs of Delivering Gas from Algeria (Hassi R'Mel) to Spain

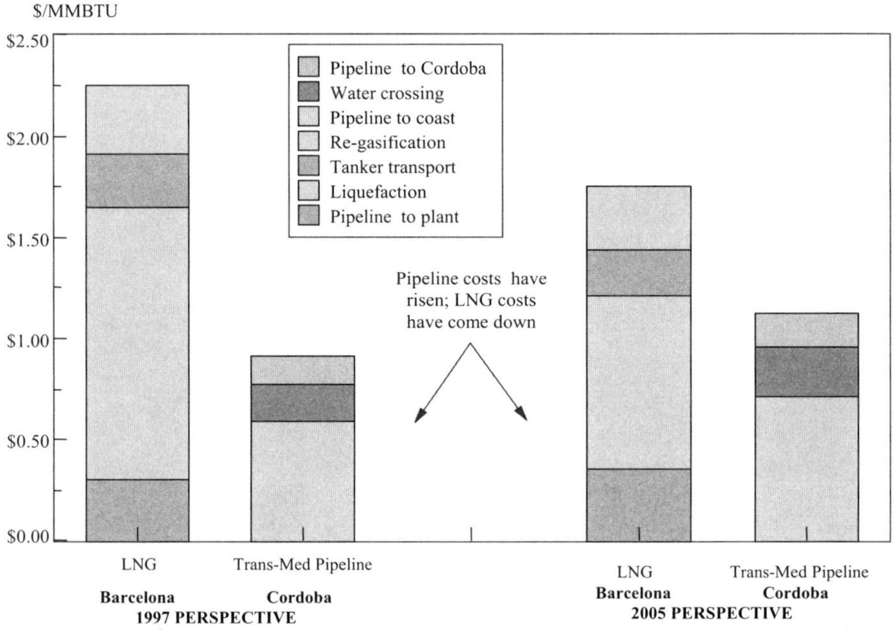

Note: Estimates assume original pipeline design sizing, two train LNG plants, present construction costs and 90% pipeline load factor operation.

Source: Author's estimates.

The Mediterranean crossing might be termed "transportation competition." The source of supply and the market destination are the same. The only question is, "Which is the cheaper way of delivering the gas?" This type of competition is comparatively uncommon. After all, a pipeline or an LNG tanker cannot serve many routes that the other system can handle. What competition does LNG offer for movements from West Siberia to Germany, or pipelines offer for movements from the Middle East to Japan?

However, the newer and increasingly common type of competition might be termed "project competition." In it, transportation is only one part of an overall package in competing supply projects. The projects are based on different gas sources and may well deliver to different destinations within the larger market. In these cases, competitive economics are complex, since they depend not only on the relative costs of transportation, but also those of competing supplies and downstream distribution within the destination market as well. Recent examples of such "project competition" include:

- Trinidad LNG versus the Mahgreb pipeline to Spain
- Algerian and Egyptian LNG to Turkey versus Russian, Iranian, Turkmen and Azeri pipelines
- Various pipeline proposals for the Indian subcontinent versus LNG
- Russian, Kazakh and Turkmen pipelines for China versus LNG
- A Sakhalin pipeline for Japan versus LNG

While the Sakhalin alternative might initially appear to be just another case of transportation competition, it differs in that the pipeline alternative as envisioned originally, requires the construction of a "Japanese gas grid" to link the various regional markets in Japan. These markets have developed around a series of isolated LNG receipt terminals and the various regions themselves are not interconnected. The construction of an integrated pipeline grid in Japan would be extremely expensive, given the high cost of pipeline construction in onshore Japan and the lack of "eminent domain" legislation in that country to facilitate acquisition of pipeline right-of-way.

The recent competition between pipeline gas from Sakhalin I with ExxonMobil as the operator, and LNG from Sakhalin II with Shell as the operator was possible because it targeted the distribution grid in the Kanto region around Tokyo, rather than attempting national distribution. Since Shell's LNG project seems to have won that particular round, the pipeline alternative may be out of the running for the foreseeable future.

These competitive projects often interact with one another within the market, making the economics of one dependent on whether or not the other goes forward. In some cases, the layout of the initial project defines the configuration of the market infrastructure, thereby shaping the course of subsequent expansions. In the Japanese case, the fact that the gas infrastructure developed in regional isolation around LNG receipt terminals clearly prejudiced the subsequent ability of Japan to accommodate pipeline supply.

LNG Cost Reduction

Both pipeline and LNG projects have been able to capitalize on this growing demand by significant cost reduction. For LNG, increases in train sizes, improved equipment design, elimination of "gold plating" and other technical improvements have brought about a substantial reduction in costs over the past decade.

The phenomenon of "gold plating" in plant design refers to an emphasis on operating reliability without particular regard to the costs involved. By building in system redundancy, suppliers were able to alleviate some of the customer concerns about supply security. With a longer history of reliable operation and with the greater destination flexibility provided by a liberalized industry structure, there is now more customer concern for costs and somewhat less for the reliability of individual plants.

However, the greatest improvement in LNG liquefaction plant costs is attributable to economies of scale. The sizes of early LNG trains were limited by the available compressors of the period. Typical trains were limited to about two million tons of liquefaction capacity. It might thus take three trains with a total of six million tons of capacity to justify a new green field LNG facility. Yet, with increasing technical improvements in compressors, train sizes have increased significantly and trains of four million tons or more are now common. Qatar has even been considering train sizes up to 7.8 million tons. Train economics respond significantly to economies of scale so that these larger sizes have been the key factor in LNG plant cost reduction.

During the decade of the 1980s, there were 17 LNG trains (both green field and expansion) built and their average size was just under 1.8 million tons per train. Figure 8.6 shows the average capital expenditure (capex) per ton (in dollars of the day) for those trains and compares it to the current cost for a 4 million ton train and a 7.5 million ton train considered by Qatargas. As is evident, most of the cost improvement over time is attributable to economies of scale. Large diameter pipelines, which require large markets, can minimize transportation costs. Marine pipelining is more expensive than onshore pipelining, but technological improvements in offshore pipelines enable them to compete in markets that were unthinkable twenty years ago.

Figure 8.6

Cost Reduction in LNG Liquefaction Plants Illustrated

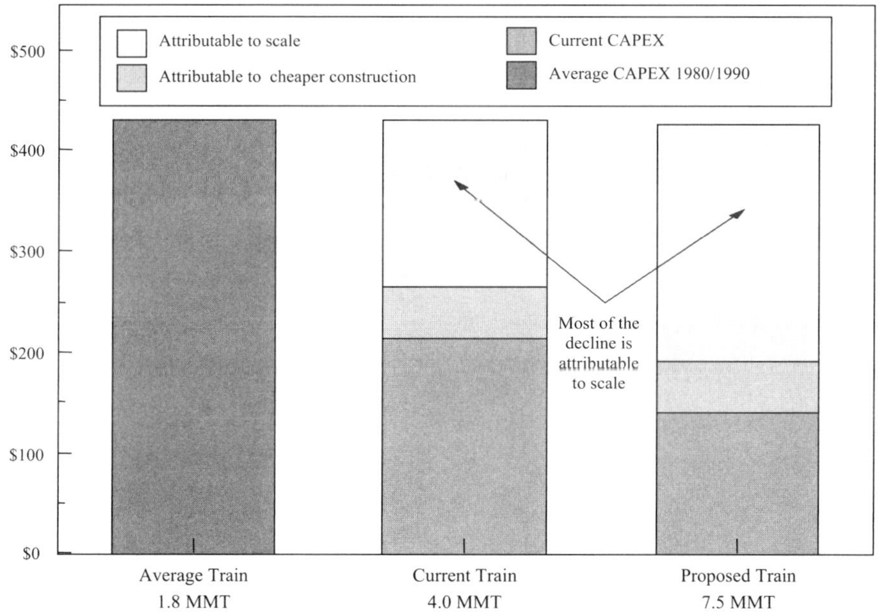

The average train size during the decade of the 1980s was 1.8 MM tons. Increases in train sizes are the major reason for the drop in costs.

Note: Basis is average cost per ton of capacity for plants built from 1980 to 1990 versus costs of current larger trains.

Source: Author's estimates.

LNG tanker transportation costs have also come down significantly. However, in contrast to liquefaction costs, very little of the reduction in tanker cost is derived from economies of scale. Early in the last decade, the most common-sized vessel had 125,000 cubic meters of capacity. More recently, sizes have increased into the 138,000 cubic meter range with the latest vessels having a capacity of over 140,000 cubic meters. As in the case of LNG trains, Qatar is considering much larger vessels of up to 250,000 cubic meter capacity. Yet the savings on economies of scale is not that great. A much greater factor is the emergence of Korean competition with the Japanese shipyards in the 1990s.

As is evident from Figure 8.7, the cost savings associated with shipyard competition outweigh the economies of scale, even for the largest LNG tankers. The unanswered questions in tanker transportation are whether the emerging Chinese interest in building LNG carriers will translate into a similar round of competitive cost-cutting and how inflationary pressures will impact on tanker construction costs.

Figure 8.7

LNG Tanker Cost Reduction Illustrated

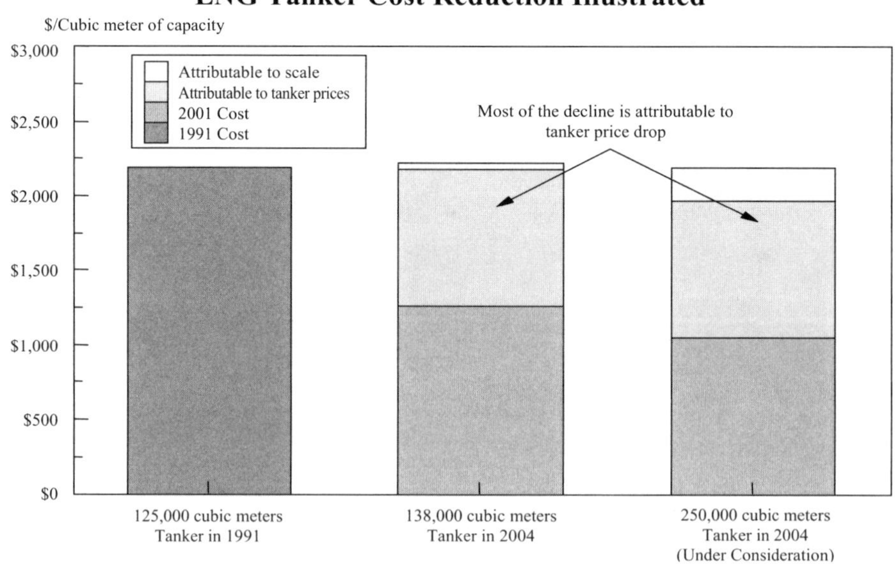

Note: Basis is 125,000 cubic meter tanker in 1991 compared with larger tankers in 2004.
Source: Author's estimates.

The pattern of LNG costs has been one of steady decline in recent years. There is, however, the possibility of a turnaround in costs that might see the recent declines reversed. The very rapid growth of interest in new LNG investments is making strong demands on the capability of qualified design-constructors, the capacity of LNG shipyards and even that to produce some types of special nickel steels. The possibility exists that the strains on the capacity to deliver these goods and services may put demand-pull inflationary pressures on LNG costs if the rapid growth continues. A possible alternative scenario is that the growth of new capacity will be slowed as customers would rather defer investments than incur rising costs for their preferred schedules.

Pipeline Economics

Pipeline economics have also benefited from new technology. The principal improvements have been in the availability of improved steels that make high pressure operation feasible and in marine pipe-laying techniques that have made deep-water crossings possible. These developments have made pipelining somewhat more competitive with LNG than it might have been ten to fifteen years ago. However, the recent inflationary pressures on materials costs appear to be more severe for pipelines than for LNG and it appears that pipelines may lose some of their relative advantages, in the absence of more technological improvements.

Earlier, the common practice was to design pipelines for operation at a maximum compressor outlet pressure of about 1,000 psi. Now, pipelines with operating pressures in the 2,000–2,500 psi range are quite common. There are two major advantages of high-pressure operation. It makes it possible to design much greater spacing between compressor stations than is possible with low pressure operation. Moreover, "line pack" is much greater at high pressures, making it much easier for the pipeline to handle transient flows in customer demands.

Gas, unlike oil, is a compressible fluid. As a result frictional flow losses in a pipeline not only reduce the pressure, but also permit the gas to expand, thus reducing the pipeline's effective carrying capacity. Therefore, gas pipeline design is especially sensitive to the spacing of compressor stations and high pressure operation makes it possible to increase the optimum distance between stations.

This a special advantage in marine pipelining, where expensive riser platforms are required to house the compressors. The combination of high pressure operation with the new deep-water pipeline technology has been especially beneficial to offshore pipelines. Figure 8.8 illustrates the improvement in marine pipeline economics with high pressures using an example representative of some of the recent North Sea lines.

Longer marine lines without intermediate compression, such as Europipe I and II, Franpipe and Zeepipe in the North Sea, or deep water lines, such as Trans-Med, Statpipe and Mahgreb, would not have been possible in an earlier period. The new Blue Stream line crossing the Black Sea from Russia to Turkey is engineered for depths of 2,150 meters (7,050 feet), and thus represents the cutting edge in pipe-laying technology.[7]

The growing interest in CCGT power generation loads can create special problems for pipeline delivery. The variability of most traditional gas loads during an operating day is relatively small. The daily concept of capacity – such as million cubic feet per day (mmcfd) in English units – implicitly assumes that transient intra-day variations in loads can readily be accommodated by pressure variations in the "line pack" gas contained within the pipeline.

However, CCGT units are often dispatched on an intermediate firing basis—often at daily capacity factors of 50 percent or less. Thus, during the limited period while they operating, their drain on line pack can be substantial and risk "running the pipeline dry." Higher pressures, increasing the effective line pack, are better able to handle this problem. The Yacheng Pipeline from offshore Hainan Island in China to Hong Kong, Asia's longest sub-sea pipeline, was specifically designed at high pressures to support power generation.[8]

Figure 8.8
Illustrative Submarine Pipelining Costs

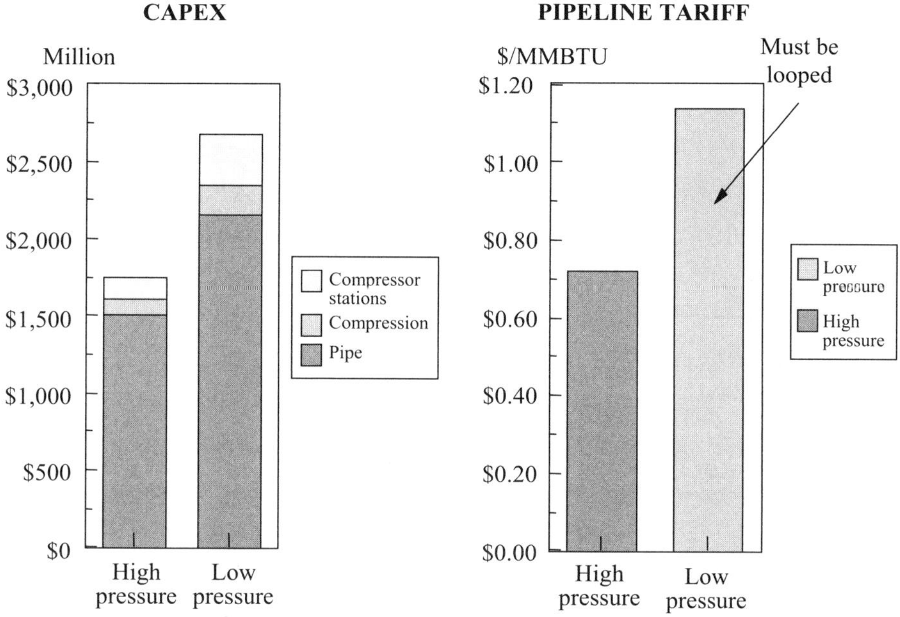

Note: Newer high pressure line compared to older low pressure line with compressor riser platforms 850 km line, 15 bcm capacity (approximately the length of the Franpipe Line in the North Sea).

Source: Author's estimates.

Factors Affecting Pipeline and LNG Competition

Economies of scale are greater for pipelines than they are for LNG. Thus, for very long haul pipeline movements, large volumes and high pressure operation are important in keeping costs under control. And since the pipeline tends to focus its deliveries on a regional market – while larger LNG plants can serve multiple destinations – market development is especially critical to large diameter pipeline feasibility. One feature that has clearly benefited pipelining has been the growth of markets, since these permit substantial scale economies.

While LNG incurs large costs for liquefaction and re-gasification, but shows much lower cost increases over distance, shorter movements tend to favor pipelines while longer ones favor LNG. This is illustrated in Figures 8.9 and 8.10, which show the scale effects for both a 1,000 km distance and a 4,000 km distance.

Figure 8.9

Delivered Cost of $0.50 Gas over 1,000 km showing the Effect of Scale Economies: High Pressure Pipelines versus LNG

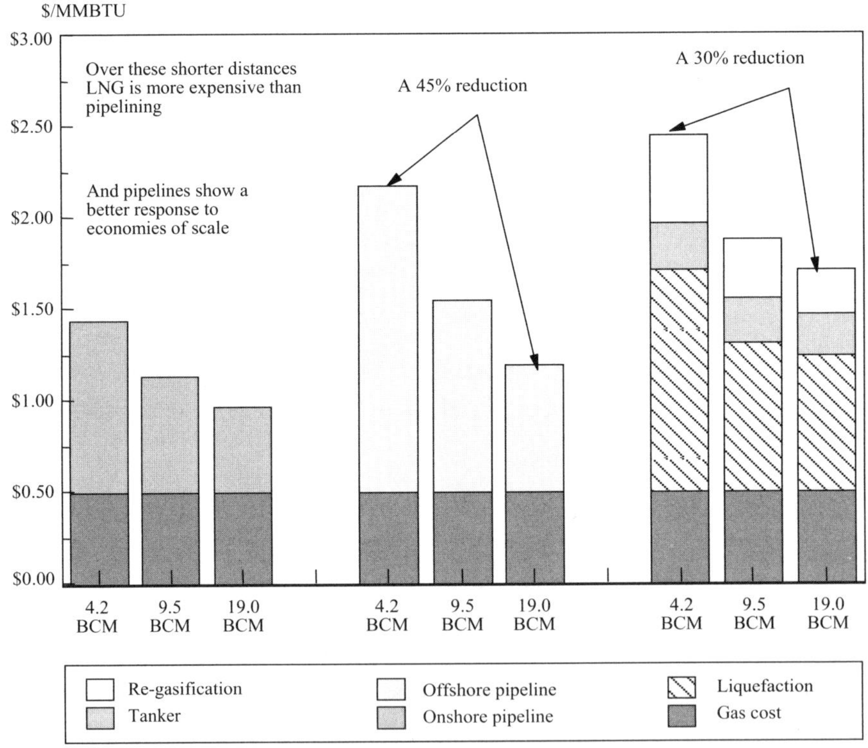

Source: Author's estimates.

Figure 8.10
Delivered Cost of $0.50 Gas over 4,000 km showing the Effect of Scale Economies: High Pressure Pipelines versus LNG

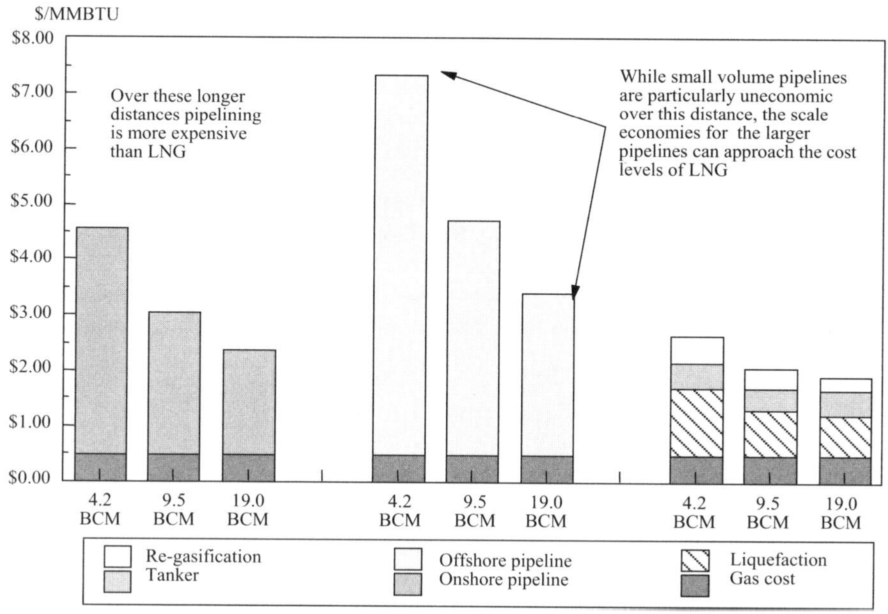

Source: Author's estimates.

Pipeline and LNG Competition: The Indian Subcontinent

The principal pipeline and LNG competition from the Middle East has involved the Indian subcontinent, Turkey and possibly Western Europe. Three alternate routes from Iran to India have been under consideration.[9] The first is an overland option transiting Pakistan on its way to India, the second is largely offshore avoiding Pakistan transit, and the third is one that stays offshore to the Iran/Pakistan border.[10] The latter option also transits Pakistan. In addition, the possibility of a pipeline from Qatar to India, such as an extension of the Dolphin line from Qatar to Dubai or the Gulf-South Asia (GUSA) pipeline proposal, has been under consideration. There has also been a proposal for a pipeline from Bangladesh in the east, but that has encountered political problems.

While these Indian pipeline projects all remain in the planning stage, LNG has already gained a substantial foothold in the Indian market. The first Indian LNG project at Dahej in Gujarat state started operation in 2004 and a second, at nearby Hazira, in early 2005. There have been a number of other proposals for LNG facilities, but the progress to commercialize them has been slower than the developers originally expected. The troubles that have plagued the Dabhol project, originally developed by Enron, together with financing problems for the power projects that would take the gas have been largely responsible for the slowdown in LNG market growth.

Figure 8.11 illustrates that the pipeline and LNG economics for a twenty billion cubic meters (bcm) delivery to India are relatively comparable (with one pipeline option significantly better) but the pipeline transit issue has heavily favored LNG up to this point. The unresolved issue of negotiating transit rights has always been a potential problem for long haul overland pipelines.

Pipelines must negotiate transit rights to move gas across third countries, raising issues of supply security and transit fees. The risk of transit interruption in the case of a dispute between the transit country and the destination country may be a concern. How the transit country will be paid for its grant of transit may also be an issue. Unfortunately, third countries are commonly in a monopoly position in the negotiation for transit rights, often making agreement difficult. While the Energy Charter Treaty is designed to deal with these issues, it has yet to prove effective. It has proved to be especially difficult to resolve the Pakistan transit issue for these two often hostile neighbors.

However, active negotiations have been under way involving Iran, Pakistan and India.[11] The discussions have been complicated by the close relationship of the United States with Pakistan in the war against terrorism and by the US policy restricting trade with Iran. If the negotiations are successful, the overland pipeline route has advantages as a supply source for Pakistan as well as for India.

Figure 8.11
Illustrative Cost Comparison for Gas Transportation: Pipeline versus LNG from Middle East to India

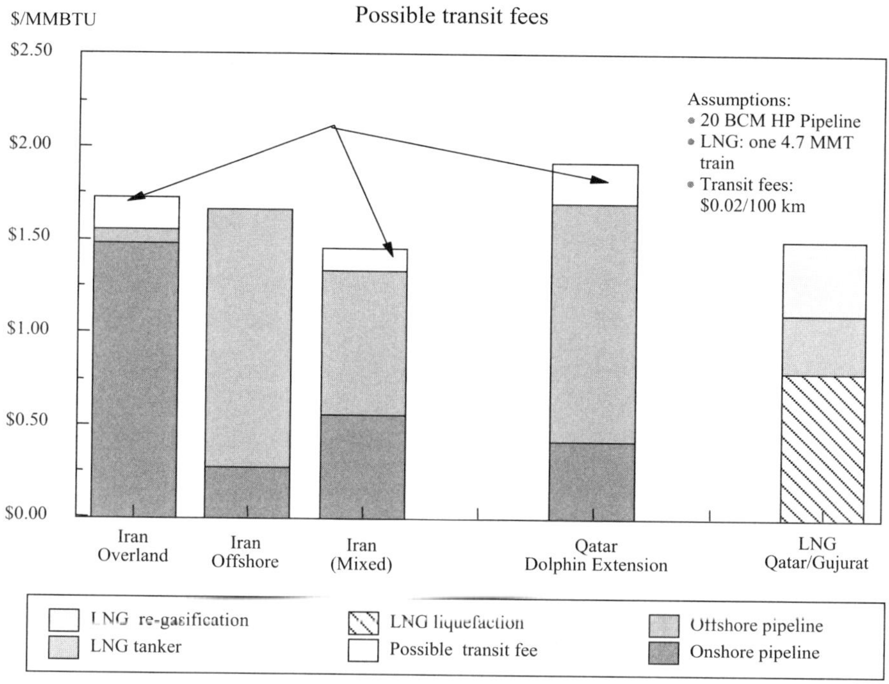

Source: Author's estimates.

It may be noted that transit fees can be paid in kind or can be paid as a fiscal charge for transit. There is no standard pattern, but Figure 8.11 assumes that the fee will be $0.02 per 100 km. This is a rate that is within the range of estimates that analysts use for such charges. In Figure 8.11, it affects the two Iranian lines that transit Pakistan. The transit estimate for the Qatar pipeline extension assumes that gas transiting the UAE would be charged a similar fee.

A pipeline from Qatar transiting the Iranian shelf would encounter one additional problem. Iran claims jurisdiction over its entire continental shelf from the beginning of the slope, thus ruling out the possibility of a shallow-water offshore line that would be outside Iranian jurisdiction. At one point, Oman was studying the possibility of a very deep-water line to India that would have avoided transit negotiations with Iran, but that plan was never implemented.

Pipeline and LNG Competition: Turkey

From Figure 8.12, it is apparent that LNG delivery costs to the Turkish market are only slightly higher than delivery via pipeline. In 2004, Turkey imported 21.2 bcm, eighty one percent of it by pipeline. The suppliers in 2004 were:

Table 8.2
Turkish Gas Imports 2004

	Pipeline (bcm)	*LNG (bcm)*	*Total (bcm)*	*Total (percent)*
Russia	14.35	–	14.35	65%
Iran	3.56	–	3.56	16%
Algeria	–	3.24	3.24	14%
Nigeria	–	1.03	1.03	5%
Total	**17.91**	**4.27**	**21.18**	**100%**

Source: Cedigaz, *Natural Gas in the World.*

In Turkey, competition may not be an issue in the near future because the country has overcommitted to new supplies.[12] Until 1994, Turkey's sole imports were delivered through a pipeline from the Former Soviet Union that transited the Balkans. In 1995, the country completed the construction of an LNG receipt terminal at Marmara Eriglisi near Istanbul and received its first LNG shipments from Algeria. It subsequently contracted for Nigerian supply and has continued to purchase short term cargoes from others. However, since the Nigerian contract, it appears to have favored pipeline supply for further expansion.

The initial pipeline competition was among pipeline imports from Iran, a proposed Trans-Caspian System from Turkmenistan and the technically innovative deep-water crossing of the Black Sea from Russia, the Blue

Stream line. The negotiations for gas from Turkmenistan were never completed and became complicated by the discovery of the large Shah Deniz gas field in Azerbaijan, much nearer Turkey and on the Trans-Caspian route. Turkey has since signed contracts with Blue Stream, which began operation in 2002, and with Iran. There is also an agreement to import Shah Deniz gas later in the decade.

As a result of its overestimate of its market growth, the Turkish gas utility Botas, has been forced to minimize its takes on existing contracts and has had to renegotiate deliveries from both Iran[13] and Russia.[14] In addition, its supply of Algerian LNG originated in the Skikda liquefaction plant, which was partially destroyed in a 2003 explosion. Turkey was willing to relieve Algeria of its delivery obligation following the accident.

Pipeline and LNG Competition: Europe

Three possible long distance pipeline routes from the Middle East to Europe involve the following lines:

- A line from Iran to Germany transiting Turkey and the Balkans
- A line from Qatar transiting Saudi Arabia, Egypt and Libya to cross the Mediterranean into Italy
- A line that transits Turkey and Greece before crossing the Adriatic into Italy.

Figure 8.12 illustrates that LNG delivery to the Italian coast (but not to Germany in the interior of Europe) are lower than any of the pipeline alternatives.

The route to Germany via the Balkans transits six countries, which is a problem that has always plagued this alternative. Such a project has been around for a long time, and in its most recent form has been termed the "Nabucco" project.[15] If it is extended all the way to Germany, it raises the problem of negotiating transit agreements with six countries. Assuming that the negotiated transit fees are at the level of $0.02 per 100 km, the transit fees fully absorb 37 percent of the total cost of transportation from the Middle East to the German border. While the Mediterranean route transits only three countries, it is longer. However, transit fees still represent 24 percent of the transportation cost. The most recent proposal for a Middle East–European pipeline project would transit Turkey and Greece before attempting a deep water crossing of the Adriatic to Italy.[16]

Figure 8.12
Illustrative Cost Comparison for Gas Transportation: Pipeline versus LNG from Middle East to Turkey and Europe

(Large diameter, high pressure designs for Europe)

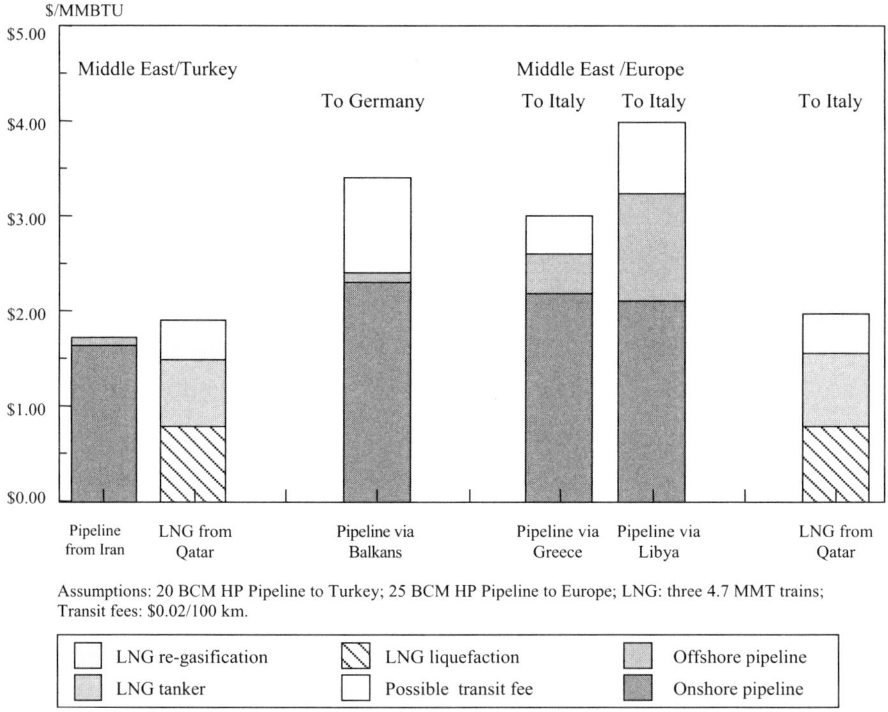

Source: Author's estimates.

At one point in the development of cross-border pipeline trade, it appeared that there might be a simple and creative solution to the transit country problem of moving gas from the Middle East to Europe. In the late 1960s, the Former Soviet Union (FSU) was facing natural gas supply problems in Georgia and Armenia in the Caucasus. Faced with the prospect of building an expensive pipeline system to bring in gas from its plentiful northern gas reserves, it opted instead for a contract for Iranian gas to be delivered to the FSU border via the Iran Gas Transmission System (IGAT-I). This pipeline began operation in 1970. The availability of Iranian supplies in the Caucasus freed up the northern gas supplies for delivery to Europe.

It soon became apparent that this was in effect an "exchange agreement" by which Iranian gas was effectively delivered to Europe via the displacement of FSU gas that would otherwise have supplied Georgia and Armenia. The displacement concept was to have been formalized through the development of the IGAT-II system in 1976. Ruhrgas in Germany, Gaz de France in France and OMV in Austria signed purchase contracts with Iran for delivery of Iranian gas to the FSU in exchange for delivery of Russian gas to Europe. The line was scheduled to be completed in 1981.

IGAT-II was actually under construction in the fall of 1978 when the first outbreak of resistance to the Shah's government occurred. In October of that year, shortly before the start of the Russian winter, the Iranian oil workers went on strike and virtually shut in Iranian oil production. Since the gas for IGAT-I was associated gas, the shipments to Georgia and Armenia were halted, causing great difficulties for the FSU in supplying its Caucasus region. In early 1989, Iranian oil workers allowed enough production to satisfy domestic needs, but refused to permit exports until the Shah abdicated.

The supply upset had a lasting effect on the concept of moving Middle East gas to Europe. IGAT-II was never completed as an export line and no one since has seriously considered using displacement for such a gas transportation movement.

The GTL Option

For longer hauls to Northern Europe, North America and Northeast Asia, GTL is the only effective alternative to LNG. GTL makes use of the

Fischer-Tropsch process, which was developed in Germany and utilized during World War II for coal-based synfuel production. The GTL processes consist of three steps:

- Reforming natural gas into a synthesis gas of hydrogen and carbon monoxide
- Synthesizing a waxy-like oil from the syngas
- Hydrocracking the syncrude into premium distillates

The process has a comparatively poor thermal efficiency and involves heavy capital expenditures, but the premium commanded by the high quality distillate may offset those disadvantages. The GTL process has two potential by-products—low level heat and fresh water from the synthesis step. The by-product water has potential value in a region that often utilizes desalination for its water supplies.

There are a number of process options. The larger scale processes, such as those developed by ConocoPhillips, ExxonMobil, Sasol and Shell, are currently the closest to full scale commercial development. However, a second group of "small footprint" processes, such as those being developed by Syntroleum and Rentech companies are in advanced stages. These processes are better adapted to smaller gas sources, such as associated gas from offshore oil fields. They may make less sense in the Middle East, where gas fields are commonly large enough to support large-scale GTL production.

As yet there are only two commercial plants in operation – Sasol's Mossgas 22,500 barrels per day facility in South Africa and Shell's 12,500 barrel per day plant at Bintulu, Malaysia. However, at least nine others[17] are in various stages of planning and development in Qatar, Nigeria and Australia. Qatar, in particular is especially active with seven plant proposals. The proposed plants represent a significant scale up from the two existing operating facilities, ranging in size from 34,000 barrels per day to 160,000 barrels per day.

Figure 8.13 compares GTL[18] and LNG economics (excluding transportation costs) with $0.80 and $0.00 feedstock gas (a case where

NGL recovery pays for field development). On account of its lower thermal efficiency, GTL is more competitive at lower feedstock costs. The lower cost of moving GTL in liquid tankers, compared to the more expensive LNG tankers, makes GTL more competitive over longer distances. Thus GTL does best when feed gas costs are low and distances are long. This is illustrated in Figure 8.14.

Figure 8.13

Cost of GTL* and LNG as a Function of the Cost of Gas

(Excluding transportation cost)

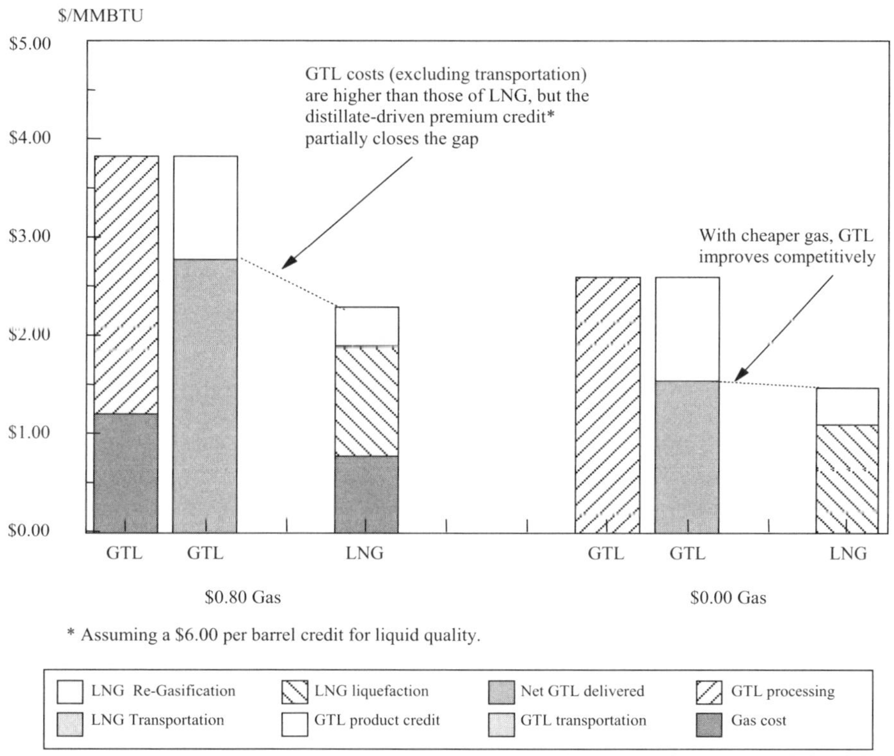

Source: Author's estimates.

Given the comparative disadvantage of GTL in markets where feedstock values are high and distances to market are short, it does not make as much sense in a supply basin that is close to major industrialized markets.

Thus GTL is obviously a more competitive option in Qatar than it would be in a place such as Trinidad.

Figure 8.14

Delivered Cost of GTL* and LNG as a Function of the Cost of Gas

(Based on a 12,000 km shipment)

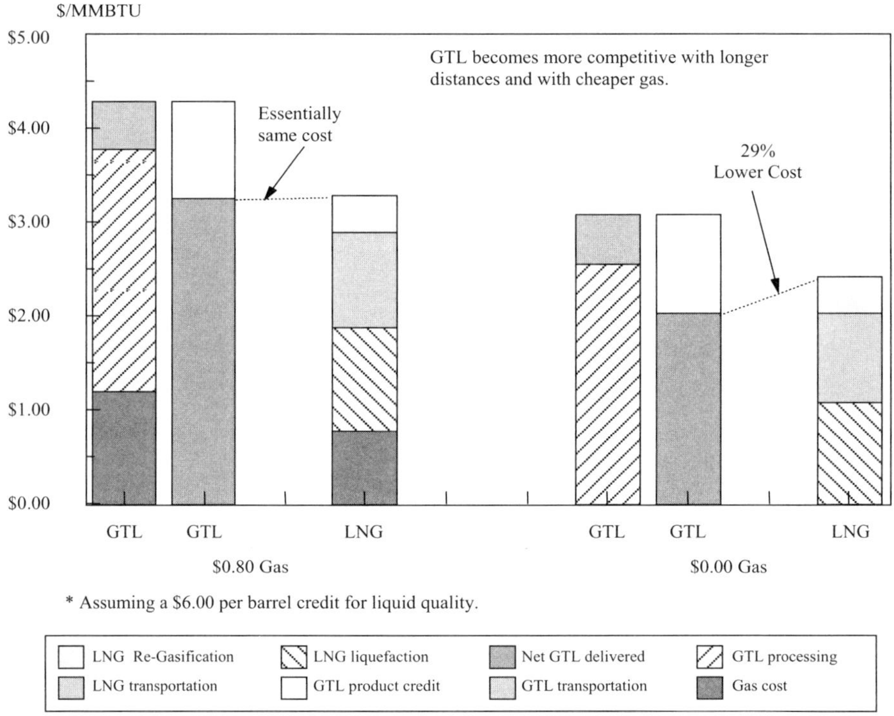

Source: Author's estimates.

Implications for the Middle East

Pipelines, LNG and GTL all have a role to play in transporting gas out of the region. Pipelines are at their best for shorter hauls and thus should dominate local and regional trade. For intermediate distances, such as for the Indian subcontinent and Turkey, pipeline growth will be inhibited by politics and competition. India will probably prefer LNG for the nearer term although successful negotiations among India, Pakistan and Iran could change that conclusion. For longer distances, such as to Europe, pipelines may potentially be competitive, but they must be large and the intermediate country transit problem is a major barrier.

Increasing LNG demand in the distant markets of North America, Europe and Northeast Asia, coupled with the Middle East's large reserves guarantee rapid growth in LNG exports to these markets. GTL is the only challenger to LNG for very long hauls but it competes in a different ultimate market. The key to GTL's success is whether the product quality premium will be enough to offset its higher costs and lower thermal efficiency.

9

Who Needs Whom? Investment in Middle East Oil

Valerie Marcel

The theme of partnerships between national and international oil companies is featured on the program of several industry conferences. At these events, both international oil companies (IOCs) and national oil companies (NOCs) profess their desire to find ways to work together. In private, however, both groups list the difficulties in finding common ground and often voice frustration at the other's expectations. Frustration on the part of the IOCs with previous joint exercises and increased confidence on the part of NOCs in their own capacity had led many on both sides to privately question the need for working with the other – at least under the present terms. Clearly, the current investment models do not allow both parties to align their respective interests.

This observation begs a first question: Do IOCs and NOCs need each other? IOCs want access to equity,[1] acceptable rates of returns, incentives for enhanced recovery and opportunities for recurring investments. However, IOCs have equity access to only 14% of the world's oil and gas reserves in countries without NOCs and to 11% of reserves in countries with NOCs present.[2] About 58% of oil and gas reserves are presently held by NOCs in countries where IOCs have only limited involvement through service contracts or technical service agreements. These terms, which are on offer in the large reserve countries, often do not meet IOC criteria for investment. As a result, a number of producers, such as Iran and Saudi

Arabia, were unable to attract bids from many big players. A Saudi Aramco executive expressed disappointment that the Saudi Gas Initiative of 2001 did not attract more investors:

> We wanted more players. It's too bad ConocoPhillips opted out of the gas initiative. It's a good company. [But they explained that] the bid didn't have the expected returns.

To bring expectations regarding returns closer in line, producers would like to see the oil majors increase their oil price assumptions to match OPEC's. Indeed, bullish or conservative price assumptions have a direct impact on the profitability estimates of future investments. Nevertheless, a number of IOCs accept less than satisfactory terms of investment for small scale or short-term investments, in order to build relationships that they hope will give them access to reserves and better returns.

IOC–NOC Relations: Attitudes and Emerging Trends

For their part, producers want capital, management skills, and technology—with various countries ranking those priorities differently and some high competence NOCs, such as Saudi Aramco, expressing no such needs. Clearly each NOC is unique, but interviews conducted with Middle East NOC managers and producing governments allow some general comments on attitudes and emerging NOC trends. Over 120 interviews were conducted during 2004 within Kuwait Petroleum Corporation (KPC), National Iranian Oil Company (NIOC), Saudi Aramco, Abu Dhabi National Oil Company (ADNOC) and Sonatrach (the Algerian state-owned energy company) as well as within the Ministries of Petroleum of these countries. One of the needs expressed was that most producers want greater access to technology, or more precisely, technology-related experience—knowing when and how to use the appropriate technology. They also need management skills to help run their operations efficiently and effectively. Some require IOC assistance to integrate sub-surface and on-field management. As a Kuwaiti Exploration and Production manager put it:

> We are better at that now, but we still need to develop that more. IOCs are good at that: integration. That's their edge. For an investment they have to calculate their return and therefore they are more efficient.

Sonatrach, for its part, needed IOCs to help it to secure important reserves in an isolated region of Algeria. IOCs have the capacity to manage large scale, technologically and logistically challenging projects. They also offer training in strategic management, risk management and financial engineering.

Furthermore, producers need investment capital when their fiscal relationship with the state is structured in such a way that their capital needs are sacrificed to government budgetary needs or when their means of revenue generation cannot meet investment requirements, such as in Iran and Algeria. Capital needs are expected to be important across the region because of the high rates of capacity utilization in a number of producing countries over 2003–04 and a slow-down of non-OPEC growth rates means that investment will be needed to increase capacity. It may be noted that Iran, Kuwait, Algeria and Saudi Arabia all have ambitious plans to increase production capacity.

Saudi Aramco, KPC and ADNOC can self-finance projects, as long as their investment capital remains insulated from short-term government budgetary needs. Others can turn to markets or to IOCs for capital. However, for lack of past experience, many NOCs do not know how to access capital markets. Also, most of them cannot open their books to market scrutiny without the state's approval—Sonatrach being an exception. They could turn to IOCs for finance, but they might not agree with IOCs on utilization rates when production capacity is raised. Indeed, all these producers plan to develop spare (idle) capacity. In Algeria, oil professionals and officials express reassuring confidence that OPEC quotas will be increased and that investors will not be left with idle capacity. Elsewhere, however, plans are to develop or increase spare capacity and if Iran and Kuwait succeed in attracting foreign investment to carry out these plans, frustration might ensue on the IOC side.

However, most large reserve holders of the Middle East will not give equity access to their reserves. Moreover, they will seek to maintain control over the management of their resources and maximize the government returns. Agreement can more easily be reached on returns, but present investment models are not sufficiently flexible to address other concerns.

Granting international oil companies the coveted right to 'book reserves' allowed Algeria and Libya to attract much needed investment and kept investors highly committed in Abu Dhabi, but equity stakes in most other large reserve holders in the region are simply excluded. For political and historical reasons, Saudi Arabia, Iran and Kuwait, which together hold 42% of global oil reserves and 54% of global NOC oil reserves, will not offer equity stakes in their oil.

Such reticence on the part of the large producers is well known, but yet another concern is less readily acknowledged by investors—the strong feeling shared by producers that it is important to maintain control of reservoir management. They are no longer simply concerned with preserving sovereignty over resources. They have been in charge for some time now and issues of sovereignty over resources are settled (unlike public sentiments, which may perceive threats to sovereignty over resources). Their first concern now is to retain control over the management of the industry—most notably control over reservoir management. A crucial problem with present models is that IOCs would control the management of reservoirs without sufficient incentives for optimizing the development of the nation's resources. This concern regarding IOCs over-producing fields arises from their own or their neighbors' past experience with IOCs during the consortia and is also fed by contemporary cases—the case of Shell's single-handed and unsuccessful management of Oman's reservoirs was cited in interviews. However, in a more fundamental sense, their concern over resource management stems from the fact that the region's professionals view their drivers as being fundamentally different from those of IOCs. In their view, the IOCs are not looking at the long-term prosperity of the country but are more concerned with the returns expected by their shareholders in the next quarter. Given these financial expectations and the limited duration of IOC involvement in the countries (10–20 years), producers often comment that it is only natural that IOCs should seek to push production for greater returns while they are in the country. This common observation regarding IOC rationale hides a lack of trust in IOCs, which is an obstacle to IOC–NOC partnerships. The producing countries in the Arabian Gulf want to see their fields developed to maximize production over 50 years or more—even if that means producing under capacity. As one professional in the region explained, "Extending the life of a field is our first priority." The following comment, made by another professional in the region, also underlines the same priority:

> Fields will last that long if you invest in them over the long run…We have important challenges ahead and we need IOCs. We want them but we also want full control over the operations. There's a perception that IOCs want to exploit our reserves. People think of the concessions.

The IOCs also want long-term engagement to guarantee field performance but it is unlikely at this stage, that these producers would welcome long-term foreign private investment. An Iranian oil professional attributed this reticence to nationalism and fear of imperialism. These political feelings regarding the development of the nation's oil resources are deeply entrenched in the societies of Arab countries and developing nations. In this respect, governments show a greater concern about repeating past history by inviting IOCs for long-term development, while NOCs demonstrate more concern for maintaining control over operations. On the latter point, the difference lies in the fact that governments are keen to maximize their oil rent and do not always agree with the NOCs' careful and slow development of resources. An official at the Algerian Ministry of Energy and Mines felt that the NOCs were not given enough incentives and capital to invest in technology and exploration. "With partners, we see that the fields produce more. Otherwise NOCs sleep on their reserves."

A senior Exploration and Production (E&P) director acknowledged a difference in views:

> The political leadership pressures the companies to develop their fields quickly and not to develop people. The government has to encourage the company to develop its people's skills and to develop its fields without a rush.

An unspoken concern on the part of NOC managers is that the government's preference for quicker generation of revenues will lead them to turn to the IOCs.

Past experience with the IOCs from the days of the Consortia in Iran, still colors views in the region's industry. Iranian oil professionals tell stories of what they feel was outrageous compensation that they had to pay the oil majors after nationalization. In view of that experience, even today, the perfect deal is considered as one that will protect the Iranian state from compensation demands. One manager involved in negotiations with IOCs explained, for instance, that IOCs wanted oil price exposure, which was denied because the Iranians were concerned that the foreign companies could claim a value for future oil in the event of a compensation claim. The Iranian negotiators also explained that the IOCs wanted security of supply, which the Iranians felt they had given with the long-term sales of oil produced through the buybacks. They also wanted access to equity and the capacity to book reserves. Regarding the last expectation, the Iranian negotiators agreed to the companies booking their share of expected oil produced, but refused to give entitlements to oil that had not been produced. Again, the concern was that in the event of an expropriation, or if the companies left because of security issues, Iran would have to compensate what "could have been" produced. A senior figure at NIOC felt that because of political feelings regarding oil and fears of imperialism and exploitation by IOCs, it is necessary to find new forms of contracts to answer NOC concerns.

Perceived attitudes among the oil majors also contribute to the resentment of the NOCs. A number of NOC managers felt that the IOCs have not appreciated the change in operation taking place in producing countries. A Russian oil professional aptly conveyed this feeling:

> Many failure stories in China [are a result of] IOC[s] overestimating what they have to offer and how valuable it is to the counterpart. I'm convinced that most of the majors don't respond properly to the changes NOCs bring to business environment. Interestingly enough, rarely NOCs are called competitors by majors, rather they see them as potential target[s] for cooperation.

In varying degrees, the industry has evolved in relative isolation (or autonomously) over the last 50 years and one oil professional explains the resulting situation:

> [T]he distance between us grew and when they were invited to return to the Middle East, they didn't change their expectations [even though things were different].

The change in effect in the nationalized industries of the Middle East is the result of a process of maturation of the NOCs. They started with little skills base, an uneducated workforce, but with large reserves and in cases like Saudi Arabia and Kuwait, with the management support of the private companies. Since then, they have expanded their core business to integrate their activities through the value chain and are now embarking on internationalization strategies where they strive to be as competitive as the IOCs.

Industry and governments must seek new investment models to respond to these changes. One model will not suit all cases. Each NOC is completely distinct and brings to the table different assets, needs and constraints. Table 9.1 illustrates this point:

Table 9.1
Profiles of Selected National Oil Companies

NOC	*Assets*	*Needs*	*Constraints*
Saudi Aramco	-Efficient -Large oil and gas reserves -Multiple grades of crude -Long-term strategic view -Investment in technology -Human resources	-Capital for refining and petrochemicals -Promote local economy -Ownership of technology -Outlets for its crude (international refineries) -Develop gas value chain	-Upstream oil closed -Political sensitivity -Domestic energy subsidies -Future rent needs of government
KPC	-Efficient refining, retail and petrochemicals business -International marketing skills -Large oil reserves	-Experience of technology -Management practices -Clarity of relations with government -Employment for nationals	-Parliamentary opposition to FDI -Heavy/Sour grade of crude -Small domestic market -Bureaucratic internal processes -Local employment quotas
Sonatrach	-Oil and gas reserves -LNG expertise -Capitalization of NOC status abroad -Geography -Relatively transparent accounting	-Management practices -Access to distant and new markets -New oil and gas reserves -Investment in technology	-Heavy labor costs -Bureaucratic internal processes
NIOC	-Large oil and gas reserves -Experience with carbonate reservoirs -Local capacity in private service companies -Geography	-Capital -Clarity of relations with government -Technology -Management practices -Investment in refining, exploration & development -Employment for nationals -Marketing skills for gas	-Sanctions -Parliamentary opposition to FDI -Domestic energy subsidies -Bureaucratic internal processes

ADNOC	-High ratio of oil and gas reserves to production and to population -Cooperative relations with foreign partners -Management processes	-Develop HR skills -Capacity to manage large projects -Ownership of technology -Investment in difficult reservoirs -Marketing capacity for products -Investment in gas (for re-injection)	-Political reticence to develop gas for export -Reliance on consultants

Source: Author's assessment.

When asked in interviews how they would define the perfect deal with IOCs, NOC managers across the region described it as a win-win deal and highlighted the needs of the IOCs. As one of them commented:

> It has to be a win-win project. We have political risk. There's the crisis with the Atomic Agency, the US embargo. [As an incentive] we can put a LT [long term] clause in our contracts.

Another manager observed as follows:

> We are both happy, both winners. It's bad when a deal favors one over another. A balance is necessary—like in a marriage.

There is almost a consensus within the companies that a perfect deal is one based on technical grounds, and not on politics. Many also thought that the deal should give acceptable rates of returns to both parties. Another NOC manager commented:

> We must pay them right, so they're happy. We shouldn't worry about the contract terms, but we should focus on the resource management.

There are interesting avenues forward that go beyond the present discussion on conventional ties between IOC and NOC, such as the conventional Production Sharing Agreements. They answer new trends that are shaping the oil and gas industry. We see, notably, an increased blurring of NOC—IOC categories. With high profile international ventures, NOCs such as Petronas are challenging the IOCs on their territory of high political risk ventures. Sonatrach, for one, is following suit with a dedicated internationalization strategy. It is no longer possible to confine "national oil companies" within their national borders. Public ownership is also becoming an elastic concept and NOCs such as Statoil and Petrobras are partially privatized but maintain a majority government stake. Most NOCs do not limit their activities to producing and selling crude, as shown by KPC which is highly integrated, maintaining a balance of approximately 1.1 to 0.9 volumes for upstream to downstream.

Both IOCs and NOCs are changing, but one side may not have fully appreciated the changes that are transforming the other side. Let us now examine some new ways for NOCs and IOCs to join forces for their mutual benefit and the issues related to these options.

1-Provision of Services

It is often said that IOCs manage risks. So one would expect to find the IOCs busy in the high-risk areas. In fact, exploration in the deep and the ultra deep acreage is now accessible to the smaller independents and NOCs (like Petrobras). The political frontier niche is getting crowded and IOCs increasingly face competition from developing world NOCs willing to take bigger political risks and lower returns, such as the China National Petrochemical Corporation (Sinopec) and China National Petroleum Corporation (CNPC).

IOCs need to strengthen their present position or find a new niche. If the IOCs are planning to open a new technological frontier and there is to be a repeat of the performance in the Gulf of Mexico, where they broke new ground in the 1990s by developing deep and then ultra-deep reservoirs, IOCs will need to invest in order to maintain their skills. The industry demographics have changed over the years and the number of skilled people available to the oil and gas industry is declining. Fewer young engineers, economists and geologists are drawn to the oil industry. It is perceived by many as an industry of the past and suffers from a negative image. Industry observers have noted that the IOCs have done little to address this problem. Universities are not feeling the pressure from industry to train young professionals for the oil and gas industry. Most producing countries have no problems attracting students and graduates to this sector, which offers the best national opportunities and enjoys a very positive image nationally, but they are acutely aware of the need to improve the industry image to maintain their market share into the future. There might therefore be room for joint action on this front.

Assuming that the oil majors commit to significant investments in research and development, give rise to a technological breakthrough and find a new niche, high costs for the medium term in the new areas mean they could be faced with the unsustainable expectations of shareholders regarding returns. Costs would also rise in the oil majors' traditional open areas in the United States and North West Europe, which are becoming more mature and where the probability of very large new discoveries is declining. They will be under pressure to replace produced reserves and to increase rates of recovery from mature areas. As Vahan Zanoyan has put it, "They face the formidable responsibility of *profitable growth* in a depletion business, where often growth alone can be a major challenge."[3]

Paradoxically, though the oil majors would seek opportunities to explore and develop new areas to beat the depletion game, they will in fact increasingly be called upon by the major reserve holders in the Middle East to assist them in managing their reservoirs and offsetting the decline of their mature fields. Projects in mature areas present bigger technological challenges and higher costs just to maintain production levels. NOCs are under similar pressures to invest to control depletion rates.

IOCs may not want this type of work. Indeed, if more significant IOC investment is not taking place in the Middle East it is at least partially attributable to the reticence of the oil majors. However, should they decide to work as a contractor for an NOC or in partnership with an NOC to develop existing fields, IOCs need to engage in a different kind of relationship with the NOC and with the government. The IOCs would need to change their mindset, their corporate culture. This is not wildcatting, and with low risk, the IOCs must accept lower returns. IOCs also have to recognize that producing countries want to control their resources. Producers want assistance, but on their own terms. The companies that recognize this are most likely to be successful with the Middle East producers. The following quote from a NIOC manager illustrates the producer's perspective:

> Service companies can provide services often at a better cost than IOCs. This is also true with Iranian service companies. They can do exploration services, seismic, drilling, tankers...There are many, many alternatives to IOCs. IOCs have to bring down their expectations and have balanced, equal relations with us. Unfortunately they are still stuck where they were 50 years ago—they still want PSAs. We decided to propose another formula, which has been able to [attract] $40bn over the last 6 years. IOCs must change their views and appreciate local capabilities, have balanced relations and reduce their expectations.

With this investment model, the problem is not so much one of access as the conditions of access. If IOCs reassess their expectations and repackage their offer and sit down to negotiate as service providers, countries like Iran may open their doors wider. At present, a share of the activities of IOCs fits into this format. In some countries it happens in haphazard forms, and there are some IOCs that may hope (some are bigger believers than others) they will someday be invited to explore and develop new fields on a turnkey basis.

The service provider model could mean increasing friction with service companies. The package IOCs offer will need to be defined in relation to the services already provided by companies such as Schlumberger and Halliburton. As it stands, large integrated projects are the domain of IOCs, except where the country is closed to foreign capital investment in the hydrocarbon sector. In this case, the service companies step in. Closed markets, such as Mexico and Iran, where foreign companies cannot lift barrels or market the oil produced, have been the domain of service companies, which operate under strict parameters of action. A recent and interesting development is the new gamut of services that the IOCs are developing to cater to the producers such as financial management tools (hedging, futures), technical consulting and systems consulting. Here again, IOCs will clash with other service providers, such banks and consulting groups who already offer these services.

On the producer side there are also choices to be made. If controlling the resources and maximizing their lifespan is truly as important to the producers as they assert, they must propose contract terms that provide incentives for optimizing their development over the long-term.

There are also broader issues to be resolved regarding the role of the NOC. With this model, NOCs can become customers that manage contractors. There is no danger here if the NOC has a strong capacity and is able to control the technological and cost parameters of the contract. They will have access to the best technology at a lower cost. However, this model is not without risks, as NOCs may lose competencies, as their skills will not be challenged by experience. As Giacomo Luciani has explained, NOCs that only partner with IOCs and have no independent activity of their own end up being sleeping partners even if on paper they have a controlling majority or appoint the senior executives.[4] It therefore becomes more difficult for the NOC to hone the skills needed to manage the development of the resources. This model may be most beneficial if applied to mature fields and complemented by other investment frameworks in other areas.

2-Partnership to Advance Sustainable Development in the Host Country

The sustainable development of the producer involves maximizing the benefits for the economy as a whole today and optimizing the development of resources for the benefit of future generations. Many producers face serious development challenges and governments count on the hydrocarbon sector to provide more than oil rent. Many of the region's oil professionals feel that IOCs care about finding mutually satisfying agreements, but that they are not as concerned about the long-term prosperity of the country. Alirio Parra explained that there is a difference between revenue maximization, which is a short-term concern, and value creation, which is long-term. From interviews, I gathered that NOCs are driven by value creation, while IOCs (and to a certain extent governments) tend to be motivated by revenue maximization. These different drivers make it more difficult to align interests in a partnership.

Furthermore, there was some discomfort within NOCs about the IOCs' social programs. Some NOC managers gave contradictory views on what the IOCs' responsibility should be in this respect. These programs challenge the government's prerogative to provide services. The IOCs get the credit and in the end, get compensated for the cost, while the NOC is challenged on its home turf. States pay for it, but are spared the hassle of providing the service to the population. One executive voiced this complaint:

> You know that IOCs bring up their rate of return with their spending on charity programs? As I told before, there is no free lunch.

Many felt that IOCs develop expensive programs, which NOCs and states could do more cheaply. These oil professionals would prefer that the funds designated for social work be given to the state to spend. After all, this money is ultimately taken out of the returns of the state as determined by the contract. Also, many NOC managers feel that IOCs do not understand domestic needs, as much as national institutions would. Many NOCs have a strong feeling of pride for having hitherto responded to the needs of the nation. However, as the state capacity to provide public services grows in the Middle East, NOCs are less called upon to build infrastructure or provide social programs. A new trend for these NOCs is the development of more classic private company-type social programs, which seek to ensure adequate development of the producing region at a lower cost, while improving the public image of the NOC.

If the corporate social responsibility (CSR) programs of private companies are to bring greater prosperity to the producing region, while overcoming the suspicion of NOCs and respecting the boundaries of state prerogatives, they must be coordinated with existing programs handled by the relevant Ministries and those put in place by the NOC. IOCs should support existing state programs with funds, but also with their knowledge and experience in the management of development projects.

In addition to benefits to the community, a partnership model might involve the NOCs gaining competencies from the joint development and application of technology. An Iranian professional commented on the potential mutual gain:

> NOCs and IOCs need each other. The NOCs offer access to reserves, while IOCs offer access to capital, expertise (the latest EOR, for example), management skills—this is a key thing missing in NOCs. We have too much bureaucracy, which brings bad management, delays…all these things disappear with an IOC. For example, if we order spare parts, the request goes to committee after committee. The IOC just gets it…And research. Iran doesn't have research centers, except the one inaugurated by NIOC—it's a shame!

In practice, however, it is difficult to transfer skills and technology. NOCs are not all well equipped to take on new skills because of weak management practices and a specific corporate culture. One manager felt that NIOC had been unsuccessful at developing "deep relations" with companies and that interest was lacking for real technology transfer. Some respondents in Kuwait said that KOC appeared to be resistant to technological transfer. Some felt the company was too risk-adverse to accept significant technological changes in their operations. Conversely, IOCs face a challenge in accessing a country with a high capacity NOC—like Aramco in Saudi Arabia. Whether working for a high or low capacity NOC, a number of national oil experts exhibit a residual resentment of foreign oil companies, an inheritance from the time of the Consortia in Iran.

> As an Iranian who went through nationalization, revolution…my view of oil [is very shaped by those events]…I worked in the consortium and was astounded to find a "no Persian speaking" sign in the managers' mess. I saw that people were accomplishing their private business in the backfield and I asked them why they were doing that. 'You should use the toilets!?' They said there were no toilets. There was no loo for the workmen, who of course were all Iranian. So now, when outsiders say that 'they will teach me something'…

It is likely that the present terms of investment are not well designed to facilitate joint development and application of technology. The buybacks in Iran or the joint study agreements in Kuwait have disappointed the partners in this respect. In the buybacks, the IOCs work with the Iranian companies and do therefore transfer technology; but they are not responsible for teaching skills and technology. KPC has joint study agreements with IOCs, through which the oil majors provide studies on specific themes, such as gas supply or geophysical acquisition, and work in a team with KPC. There is skills transfer, though it appears to be limited, according to an IOC executive. In Sonatrach, the mood is different and the NOC appears very keen to learn from the IOCs, perhaps because Sonatrach has been put to the test and is exposed to competition in its domestic upstream business.

There is no easy formula for a partnership that allows NOCs to *control* the development of resources while they *learn* the skills of IOCs. Terms can be found, but the will must be there. The crucial issue for NOCs is trust and the lack of it is a serious obstacle to developing IOC–NOC partnerships. This legacy will need to be overcome to meet the industry's investment challenge over the next 10–20 years. New contractual arrangements may be needed to define new types of relationships between producers and IOCs. Breaking with the legacy of the past includes rethinking equity-based contracts—especially in Iran.

For this, private companies must be evaluated on more than their reserves by shareholders and financial analysts. Building partnerships will also require some time-tested human skills. Partnerships can be built on chemistry between people and their relationships. On the IOC side, it requires cultural sensitivity and good listening. IOCs should not underestimate the knowledge of NOCs. After all, the Middle East NOCs have kept their industry running with little help from the IOCs for the past 30 years. In fact, the Iraqi oil professionals who kept the oil flowing during 12 years of UN sanctions gave a dramatic illustration of ingenious self-reliance.

For their part, IOCs need to know who decides, who regulates, what committees they should refer to and who is on those committees? IOCs also need a clear legal framework—which they don't find often enough. Being integrated companies, IOCs want access to a value chain so they can give the state maximum benefits throughout the chain.

3-Joint Ventures between IOCs, NOCs and Private Local Companies

Another innovative partnership model is to have an IOC partner with the NOC and a local private company. Such partnerships have successfully been applied in the petrochemical and refining sectors and could be developed for upstream and midstream development. A model for such partnerships is the Equate joint venture, successfully developed in Kuwait. This involved PIC, the petrochemicals subsidiary of KPC, joining forces with Dow Petrochemicals (and initially including Union Carbide), as well as local private companies (which took up a 10% share). The company is independent from the state and from the NOC and functions like any other commercial entity. Its international partners have instilled the company with efficient, Western-style management practices as well as technical expertise. Some of Equate's good practices have been learned by the NOC partner and applied in the PIC subsidiary—notably PIC has changed its management, information technology as well as health, safety and environmental processes (HSE).

As for the application of this model to the upstream, there is a question as to whether states in the region would welcome private domestic equity any more than they welcome foreign oil companies. The participation of the private sector in the hydrocarbon sector has historically been limited in the region. In Kuwait, Jill Crystal describes the historical pact between the rulers and the merchant class whereby the oil industry is state-controlled and the ruling Al-Sabah family agreed to stay out of Kuwaiti business. The merchants have access to lucrative contracts, preferential monopolies and dealerships and receive their share of the oil revenue windfall.[5] It is unlikely that states in the Middle East would relinquish their control over the majority of their hydrocarbon resources, as they derive their power from their monopoly over the generation and distribution of oil revenues. However, some minor hydrocarbon assets could be shared along the lines of a NOC–IOC–local private company partnership, with potential gains for all partners. From the government's perspective, such a deal would reinforce national policies by giving new opportunities to the private sector, while maintaining control over the development of the resource through the NOC's involvement, and perhaps masking slightly from public opinion the role of foreign interests in the investment.

4-Joint Ventures Abroad

Joint ventures between IOCs and NOCs in a third country offer a neutral territory in which to develop relations. Here the NOC is learning management practices, application of technology and project management from the industry leaders—provided it takes a large enough stake to get a seat on the board. The NOC is also in a more receptive mode, being challenged by a new environment, and not intent on protecting national sovereignty. The NOC and the IOC partners can develop personal and institutional relationships through such deals that would contribute to overcome trust issues on the part of NOCs. Through a demonstration of the assets and skills of NOCs, IOCs can also learn the value of their partners. To a certain extent, this was the experience of Petrobras, the Brazilian NOC. Petrobras sought to pool resources with IOCs through international ventures. According to one company executive, this allowed the partners to meld together each partner's experience and technology.

Though joint ventures in a third country can help build a relationship between IOC and NOC that facilitates future deals, they cannot realistically be considered a direct avenue to investment in the partner NOC's home turf, nor can the IOC be expected to risk its reputation on an alliance with a weak partner. Joint ventures abroad must be beneficial to the IOC partners irrespective of potential investment gains in the NOC-partner's home territory. Joint ventures (JVs) do not survive long when they are based on hypothetical gains. As a result, the lower capacity NOC will find it difficult to attract an IOC into a partnership abroad. International private companies will seek out partners that bring value to the table.

What assets can NOCs contribute? NOCs themselves are only beginning to identify and capitalize on their assets. Most NOCs from the Middle East do not want to play on their national status to win bids outside the country—even in countries where they recognize that in doing so, they may have an advantage. There is a marked preference among NOCs for playing on the same level field as the other private groups and demonstrating their equal competence. NOCs want to be seen by other NOCs and governments of host countries as international, not national.

However, having said that, Sonatrach is quite deliberate in capitalizing on its experience with dealing with the social environment when approaching host countries—and some peers recognized this as an Algerian advantage. Some Algerian oil professionals felt it was about respecting other cultures in foreign ventures. In this respect, they explained that in some cases Sonatrach had more to offer the host countries than oil majors because of its capacity to listen, its common values and experience, and its capacity to "share the fruit of [their] longer experience in the industry." Another professional commented:

> The advantage of an NOC is that we make others more serene about the fact that we will listen to their needs and understand them. There is a cultural proximity between NOCs and host countries. Also that we will not be too greedy – [they recognize] this is our reputation.

The NOC partner can also support the negotiation and entry in a hard-to-access country. A senior international downstream manager in the Arabian Gulf explained: "We use the label to our advantage. The label implies we give a guarantee of access to our crude." This is particularly useful in Asia, where countries are largely dependent on Middle East supply. More concretely, NOCs can also help deal with unions. They can also understand how to work through a bureaucratic system, such as in China.

There is awareness among the NOCs in the region of the advantages and disadvantages of being associated with their national flag. Kuwait's strong relationships with Arab states (supported by a tradition of lending money to neighbors) are an advantage, where IOCs are viewed with suspicion. Being the NOC of an Islamic country is also an advantage for companies investing in the Islamic world. As one NOC manager put it:

> It's true that being an NOC may be the cherry on top that gets the deals in Iraq or in Yemen. We do have an intimate understanding of Muslim countries.

However, it is noteworthy that for Kuwait Petroleum International (KPI) operating in Europe, the affiliation with a Muslim, Arab country was perceived as a potential threat to their market. As a result, they re-branded there as Q8. The company's studies showed that the "Q8 sails" image was of a youthful, vibrant brand, and not a state brand, and according to one executive, they want to keep it that way. Their governments also may support their NOC in its negotiations. However, this NOC advantage is more relevant for access to markets than for exploration and development of a field, because the host government will want the best contractors for the latter task.

NOC managers are acutely aware that while they boast of these NOC advantages abroad, they have a credibility problem to overcome in the perceptions of their potential partners as well as their own employees. One Algerian professional explained, "There is a perception of risk in doing business with NOCs that is not the same with IOCs (in particular with oil supermajors)." Sonatrach needs a successful outcome in its international ventures to counter this perception and increase its credibility among peers.

Middle East NOCs bring more assets to downstream joint ventures than in the upstream, where the operator's responsibility is great and a strong track record of applying the right technology and controlling costs is crucial. Because NOCs have less experience with the management of large projects, cost control and new geological conditions, they have a great deal to learn from IOCs in these joint ventures abroad. In terms of international downstream ventures, NOCs like KPC, Sonatrach and Saudi Aramco have already integrated their activities significantly and have a presence in refining, but they can learn from the IOCs' management practices and marketing skills.

5-International Strategic Alliances throughout the Supply Chain

As a number of NOCs expand the scope of their activities through the value chain and internationally, they seek to attract foreign investment in new core business and to acquire assets abroad to give them better access to markets and achieve further integration. There are, in this respect, opportunities for international strategic alliances between IOCs and NOCs, with each party leveraging its assets to access the missing link in its supply chain.

A number of the traditional petroleum producers are expanding the focus of their core business and are turning increasingly to gas to support oil exports in generating rent as well as to allow greater integration of activities. There is potential for joint gas development in these producing countries where the hydrocarbon industry was otherwise closed to foreign investment. Gas is indeed a less politically sensitive resource than oil, as indicated by the Saudi Gas Initiative. However, when gas products are planned for domestic consumption, subsidies and government fixed prices often limit the attractiveness for investors (including for the NOCs) as is the case with refining. For some time, large gas reserves holders in the Middle East (with the exception of Qatar) saw greater value in gas as a substitute for domestic oil consumption (than in gas as a commodity for export) because it would free lucrative petroleum for export. This attitude is changing. Faced with declining petroleum reserves in Algeria and new technological capacity in gas development, Sonatrach and Iran want to be the gas hubs of the future, accessing markets left open by Qatar and Russia. In terms of domestic supply of energy, the scene is also changing in the major producers. The Algerian and Iranian governments had submitted bills to Parliament aimed at reforming the domestic subsidy system, and the Saudi government is expected to remove its subsidy system in conformity with World Trade Organization (WTO) regulations.

Likewise, there is potential for foreign investment to support greater integration of NOC activities. A Saudi Aramco executive explained it thus:

> What would be of interest is a gas value chain. IOCs bring capital and know how. It would increase opportunities for privatization—which would decrease the funding burden of the government and bring the IOCs in for power generation and the other key needs of the kingdom.

Sonatrach is looking for new "forms of penetration" with IOCs. Thinking beyond the JV format, the company seeks to engage IOCs differently, with asset swaps, for instance. This is an interesting way to develop new skills and to access new markets. Sonatrach is discussing an asset swap with Statoil. The companies will exchange stakes of the same value in each other's gas fields (the Christina field against an undetermined Algerian field). Finding stakes of equivalent value with similar tax regimes is a particular challenge in these ventures. The companies are discussing exchanging a minority stake in the order of 20–30%. It is important for Sonatrach to secure a large enough stake to have a seat on the Board. The recent agreement between BP and Sonatrach to jointly book capacity at the LNG re-gasification terminal at Isle of Grain in Kent is a successful joint venture which has added value to both parties. These deals are signs of a new trend of increasing collaboration in the European gas supply chain.[6] There are also potential complementarities between gas producers, such as Sonatrach, and downstream gas marketers, both concerned about demand security. As Keith Myers and Philippe Carpentier have argued, supply chain joint ventures can create greater flexibility for both ends of the chain, while providing greater supply and demand security. Companies like Sonatrach may also find access to new markets in Europe easier with a local partner, as European market liberalization places new conditions on suppliers.

In terms of forward integration, NOCs also seek to pursue international refining and marketing, transportation and petrochemicals. Saudi ventures in the United States and Japan are significant in this respect. These partnerships offer the large exporter of crude security of markets – which is of particular importance to Saudi Aramco – and offer the foreign partner a way of reducing the capital employed in the low-return downstream business—without exiting it completely.

NOC objectives and methods in international refining are not necessarily compatible with IOCs. Their international refineries tend to be subject to limitations regarding the source of the crude—as they are outlets for the national crude. This is changing in Kuwait and elsewhere, and we might expect oil to be treated increasingly as a fungible commodity by producers as the state expectations grow regarding the refineries' bottom line and as NOCs increasingly sign up for joint venture refineries with IOCs and the dominant producer culture of price and destination control gives way to more open market practices. The CEO of a leading NOC noted that the volumes and share of transported and traded energy is always increasing and this common industry concern could be the basis for a new business model linking NOCs and IOCs. Partnering of IOCs and NOCs may also help both parties enter a 'closed' market, as was hoped with the Saudi Aramco–ExxonMobil joint venture to access the Chinese downstream market.

These various partnerships will not necessarily give IOCs equity access to petroleum reserves on the NOC's home turf. However, they do offer benefits to the IOCs if they build relations that promise future access to reserves and if there is a strategic gain in the immediate term. Also, to the extent that these partnerships are based on joint decision-making and operations, there is a real opportunity for transfer of technology and management practices to the benefit of the NOCs.

6-NOC buys IOC

NOC ambitions may lead to a new phase of industrial acquisitions, where the NOC purchases smaller private oil companies in a drive to acquire the skills, the technology and the international exposure that it lacks. This option becomes more viable in the current period of high oil prices thanks to the increased revenue windfall for producers—though assets could not be bought cheaply until prices fell. An oil supermajor would be too large a bite for even the largest NOCs. The capital commitments could not be justified to the state, which invariably has other plans for these revenues. Also, it is likely that the IOC's government would react to its flagship company being bought by an OPEC producer. Indeed, in this case, political sensitivity in reaction to foreign investment may not come from the producer, but from the IOCs' home country. When the Kuwait Investment Authority (KIO) bought a 23% stake in BP in 1987–88, BP successfully appealed to the British government to intervene. BP board members argued to the British Government that what was ostensibly a portfolio transaction by KIO was in fact a veiled attempt on the part of KPC to take over BP.[7]

Less controversial and more financially viable, an NOC might acquire an independent. Following a strategic assessment of the NOC's needs and capabilities, the acquisition of a company with the needed technology and experience might fill the necessary skills gap for the NOC. For example,

Kuwait Oil Company (KOC) lacks technology and experience with heavy crude, which is set to take a greater share of its petroleum production. Meanwhile, Kuwait's fields with lighter crude will present ever-greater challenges as they mature and face decline. It therefore needs to hone its skills for enhanced recovery in its existing fields on the one hand, and for the production and development of heavy crude on the other. It could seek to acquire an independent with extensive experience in heavy crude or a mid-cap international oil company with combined experience of heavy crude and mature fields. Saudi Aramco, for its part, needs to secure new refining outlets for its crude and might benefit from the extensive downstream network of a mid-cap oil company.

The challenge with such acquisitions comes from the clash in corporate cultures. These private oil companies are attractive to the buyer notably because of their lean operations and flexible management processes, and yet, in a merger, these qualities are unlikely to survive in the NOC's operating environment. It is therefore necessary to keep the private company at arms' length and largely independent from the NOC, while devising mechanisms and processes for exchanging views and developing joint strategies.

Conclusion

Successful new business models will depend on a careful alignment of each party's objectives, needs and assets. This chapter has suggested potential new ways for NOCs and IOCs to join forces for their mutual benefit. The difficult task lies in finding a venture where both parties bring complimentary assets to the table. Another obstacle, as we have seen, relates to how IOCs and NOCs perceive each other. Neither seems to feel 'understood' by the other, as suspicion on the NOC side or frustration on the part of IOCs often mar negotiation efforts. Nor do they fully appreciate how the other has changed. NOCs are unlikely to give turnkey contracts as they want to control and participate in the development.

This change in NOC expectation means IOCs must either work for the NOC or work with it. By developing new business models, as discussed here, they will become partners. Successful partnerships will lead the NOCs to gain new competencies and increasingly they will be competitors.

10

Global Energy Outlook: Implications for Growth and Investment Prospects in the Middle East Petroleum Sector

Fatih Birol

This chapter outlines the implications of the world's relentless appetite for energy on growth and investment prospects in the Middle East petroleum sector. It draws on the main findings of two reports from the International Energy Agency, the *World Energy Outlook 2004* and the *World Energy Investment Outlook, 2003*.

The IEA's above-mentioned energy projections conclude that if governments adhere to the policies in force as of mid-2004, the world's energy needs in 2030 are expected to be almost 60% higher than at present. Fossil fuels will continue to dominate the global energy mix, meeting some 85% of the increase in overall energy use. Although the earth's energy resources are more than adequate to meet this demand, many challenges will have to be overcome including the procurement of sufficient investment to turn these resources into production in a timely manner. Given the huge oil and gas resource base in the Middle East,[1] the world is expected to become progressively more dependent on energy supplies from the region. Although exploration and development costs in the region are very low, financing may be a constraint, particularly in those countries where government policies or perceived geopolitical factors prevent or discourage foreign involvement.

Global Energy Demand

Primary Fuel Mix

World primary energy demand is projected in the *World Energy Outlook 2004* Reference Scenario to expand at an average annual increase of 1.7% per year between 2002 and 2030 (see Figure 10.1). Demand will reach 16.5 billion tons of oil equivalent (toe) in 2030 compared to 10.3 billion toe in 2002. The projected rate of growth is, nevertheless, slower than over the past three decades, when demand grew by 2% per year. Fossil fuels will continue to dominate global energy use. The share of renewable energy sources will remain flat, at around 14%, while that of nuclear power will drop from 7% to 5%.

Figure 10.1

World Primary Energy Demand by Fuel

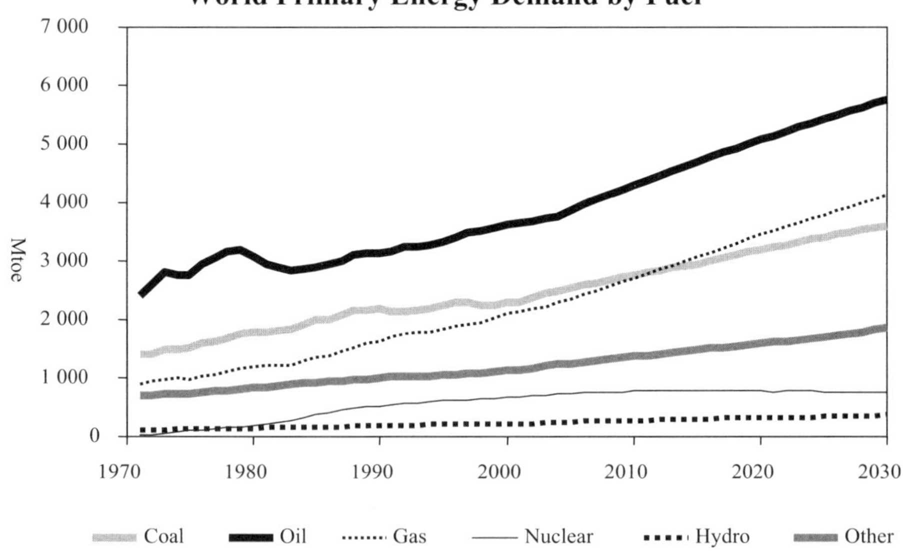

Source: *World Energy Outlook 2004*, IEA.

Demand for oil is projected to grow by 1.6% per year, from about 82 mbpd in 2002 to 90 mbpd in 2010 and 121 mbpd in 2030. Oil use will become increasingly concentrated in the transport sector, which will account for two-thirds of the increase in total oil use. Transport will use 54% of the world's oil in 2030, compared to 47% now and 33% in 1971. Oil will face little competition from other fuels in road, sea and air transportation during the projection period. In OECD countries, the use of oil in the residential and services sector will decline sharply. In non-OECD countries, transport will also be the main driver of oil demand, though the industrial, residential and services sectors will also see steady oil-demand growth. In many developing countries, oil will remain the leading source of modern commercial energy for cooking and heating, especially in rural areas.

Primary demand for natural gas will grow at a steady rate of 2.3% per year over the projection period. By 2030, gas consumption will be about 90% higher than now, and gas will have overtaken coal as the world's second-largest energy source. The share of gas in total primary energy use will increase from 21% in 2002 to 25% in 2030. The power sector will account for 60% of the increase in gas demand, with its share of the world gas market rising from 36% in 2002 to 47% in 2030. The power sector will be the main driver of demand in all regions. This trend will be particularly marked in developing countries, where electricity demand is expected to rise most rapidly. Natural gas will remain the most competitive fuel in new power stations in most parts of the world, as it is the preferred fuel for high-efficiency combined-cycle gas turbines (CCGTs). A small but growing share of natural gas demand will come from gas-to-liquids (GTL) plants and from the production of hydrogen for fuel cells.

Coal use worldwide is projected to increase by 1.5% per year between 2002 and 2030. By the end of the *Outlook* period, coal demand, at just over 7 billion tons, will be about 50% higher than at present. The share of coal in total primary energy demand will, nonetheless, fall slightly, from 23% to 22%. China and India, both of which have ample coal supplies, will account for more than two-thirds of the increase in global coal use over the projection period. Power stations will absorb most of the increase in coal demand, though coal will continue to lose market share in power generation in all OECD regions and in some developing regions. Coal consumption will increase slowly in end-use sectors. Industry, households and services in non-OECD countries will use more coal, more than offsetting a continuing decline in final consumption in the OECD countries.

The role of nuclear power will decline progressively over the *Outlook* period. The construction rate of new reactors is expected to keep pace with the retirement rate of old reactors. The reasons are twofold: first, nuclear power will have trouble competing with other technologies and second, many countries have restrictions on new construction or policies to phase out nuclear power. As a result, nuclear production will peak soon after 2010 and then decline gradually. Its share of world primary demand will drop from 7% at present to 6% in 2010 and to 5% by 2030. Nuclear output will increase only in a few countries, mostly in Asia. It is projected to fall in Europe. These projections are, however, subject to considerable uncertainty. Possible changes in government policies and public attitudes with regard to nuclear power could allow it to play a much more important role.

Hydropower production will expand by 1.8% per year over the projection period, a slightly faster rate than that of global primary energy demand. The share of hydropower in global electricity generation will nonetheless drop—from 16% in 2002 to 13% in 2030. Most of the increase in output will occur in developing countries, where there are still considerable unexploited resources and where public opposition is less of an impediment to new projects.

The role of biomass and waste, the use of which is concentrated in developing countries, will gradually diminish over the projection period. Globally, its share of primary energy demand will drop from 11% in 2002 to 10% in 2030, as it is replaced with modern fuels. In absolute terms, the consumption of traditional biomass and waste in developing countries will continue to grow, but the growth rate will slow down over the projection period.

Other renewables – which include geothermal, solar, wind, tidal and wave energy – will grow faster than any other primary energy source, at an average rate of 5.7% per year over the projection period. However, the 2% share of "other renewables" in world demand would still be small in 2030, compared with 1% share in 2002, because they start from a very low base. Most of the increase in the use of such renewables will be in the power sector. Their share in total electricity generation will grow from 1% in 2002 to 4% in 2030. Most of the increase will occur in OECD countries, many of which have policies to encourage the adoption of new renewable-energy technologies.

Regional Trends

Two-thirds of the increase in world primary energy demand between 2002 and 2030 will come from the developing countries. OECD countries will account for 26% and the transition economies for the remaining 8% of this demand. Consequently, the current 52% share of the OECD countries in world demand will decline to 43% in 2030, while that of the developing countries will increase, from 37% to 48%. The share of the transition economies will fall from 10% to 9%. The increase in the share of the developing regions in world energy demand results from their more rapid economic and population growth. Industrialization and urbanization will also boost demand. More people will live in towns and cities and will be better able to gain access to energy services. Increases in real prices to final consumers, a result of the gradual reduction in subsidies and rising international prices, are not expected to affect energy-demand growth in developing countries very much.

The share of developing countries in global demand will increase for all primary energy sources, except non-hydro renewables. The increase will be most pronounced for nuclear power, where production will fall in the OECD countries while expanding in China and other parts of Asia. The developing countries will account for 18% of world nuclear output in 2030, compared with only 4% in 2002. Their share of world coal consumption also increases sharply, from 46% to 61%, mainly because of booming demand in China and India. Developing countries will account for almost two-thirds of the 43 million barrels per day (mbpd) increase in global oil consumption between 2002 and 2030.

Despite relatively strong growth in per capita energy use in developing regions, consumption will remain much lower than in the rest of the world. By 2030, per capita primary energy consumption will average a mere 1.2 toe in developing regions, compared with 5.4 toe in the OECD countries and 4.7 toe in the transition economies.

Global Energy Production

Resource Availability and Production Prospects

The world's energy resources are adequate to meet the projected increase in energy demand until 2030 and beyond. In other words, a supply crunch is not expected for at least the next three decades, though short-lived bottlenecks could constrain supply of a given fuel at a given time and drive up prices. At issue, however, is how much it will cost to develop and transport fossil-fuel and other resources to meet demand.

Proven reserves of gas and coal far exceed the cumulative amounts of both fuels that will be consumed over 2002–2030, and more reserves will certainly be added during that time. Proven conventional oil reserves today are sufficient to cover all the oil that will be needed until 2030. However, additions to reserves from new discoveries and from "proving up" probable and possible reserves will be needed if production is not to peak before then. Non-conventional crude oil reserves are, in any case, very large and will replace any shortfall in conventional supplies as prices rise. There is also considerable potential for expanding oil production from gas-to-liquids plants, drawing on natural gas resources that would otherwise not find a market. Reserves of uranium for nuclear power production are also plentiful, while the technical potential for renewable energy sources, in principle, may be considered almost limitless.

The geographical sources of incremental energy supplies will shift markedly over the projection period, mainly in response to cost factors and the location of resources. From 2002 to 2030, more than 95% of production increase will occur in non-OECD countries, against about 70% from 1971 to 2002. It may be noted that most low-cost fossil-fuel resources are located in non-OECD countries.

OPEC countries, especially in the Middle East, will expand their production of oil most rapidly over the projection period. Their resources are large and their production costs are generally very low. Their share in total production will surge from 37% at present to 53% in 2030. Over the rest of the current decade, however, non-OPEC countries are expected to contribute half of the increase in global production. World oil production will not peak before 2030, although output in some regions will already be in decline by that time.

The biggest increases in natural gas production are projected to occur in the transition economies and the Middle East. Africa and Latin America will experience the fastest rates of increase. The cost of non-associated gas production is believed to be lowest in the Middle East. Production costs are rising in the most mature producing regions, particularly in North America and Europe.

China will reinforce its position as the world's leading coal producer, accounting for around half the increase in global output over the projection period. The United States, India and Australia will remain the next biggest producers. Coal production in Europe will continue to decline as subsidies are reduced and uncompetitive mines are closed.

Energy Investment

The IEA's *World Energy Investment Outlook* of 2003 assessed, fuel by fuel and region by region, the prospects for and possible barriers to investment in the global energy sector through to 2030. It concludes that worldwide investment requirements for energy-supply infrastructure over the period 2001–2030 are $16 trillion, or $550 billion a year. This investment is needed to replace existing and future supply facilities that will be exhausted or become obsolete during the projection period. It is also needed to expand supplies to meet the projected primary energy demand growth. Capital needs will grow steadily through the projection period. The average annual rate of investment is projected to rise from around $450 billion in the current decade to $630 billion in the 2020s (see Table 10.1). Estimated energy investment was $410 billion in 2000.

Table 10.1
World Energy Investment Requirements
($ billion in year 2000 dollars)

	2000	*2001-2010*	*2011-2020*	*2021-2030*	*Total 2001-2030*	*Share of total 2001-2030 (%)*
Oil	87	916	1,045	1,136	3,096	19
Gas	80	948	1,041	1,157	3,145	19
Coal	11	125	129	144	398	2
Electricity	235	2,562	3,396	3,883	9,841	60
Total	**413**	**4,551**	**5,611**	**6,320**	**16,480**	**100**

Source: *World Energy Investment Outlook, 2003*, IEA

Although the total investment needs are large in absolute terms, they are modest relative to the size of the world economy, amounting to only about 1% of global GDP on average over the next thirty years. The proportion is expected to fall slightly over the projection period, from 1.1% in the current decade to 0.9% in the 2020s. However, needs differ sharply among regions, ranging from half a percent of GDP in the OECD countries to 5% in Russia.

Several factors could discourage, prevent or delay investment in particular regions or sectors. Prices will have to be high and stable enough to ensure adequate returns. Access to reserves, fiscal and commercial terms, the evolution of environmental policies and technological developments will also be key factors in future investment. There is considerable uncertainty about the amount of energy resources in certain regions and about how much it will cost to exploit them.

Implications for the Middle East Petroleum Sector

Given its huge resource base, the level of investment in expanding oil and gas production capacity in the Middle East will be vital to global energy market prospects in the medium to long term. Mobilizing that investment will depend largely on the production and investment policies of the key producers. Although the costs of developing the region's vast reserves are lower than anywhere else in the world, restrictions on foreign involvement in many countries and the dependence on national oil companies for a large share of state revenues might constrain the amount of capital available for investment in increasing production capacity.

The Middle East Oil Sector

Over the longer term, Middle East countries will account for most of the increase in global oil production. Of the projected 31 mbpd rise in world oil demand between 2010 and 2030, around 29 mbpd will come from OPEC countries in the Middle East[2] (see Figure 10.2). This region holds about 60% of the world's proven oil reserves. Production costs there are among the lowest anywhere in the world, averaging less than $2 per barrel in the Arab Gulf countries. Investment costs are also very low, at less than $5,000 per barrel per day of capacity. Saudi Arabia, Iraq and Iran are likely to contribute most of the increase in Middle East production. Saudi Arabia will continue to play a vital role in the global oil market balance. Its willingness and ability to make timely investments in oil production capacity will be a major determinant of future price trends. Saudi Arabia produced 9.5 mbpd in August 2004. There are some 15 fields in production, including a large number of super-giant fields. Seven of these have a combined production of 7.25 mbpd of capacity: Ghawar, Abqaiq, Shaybah, Safaniyah, Zuluf, Berri, and Marjan. Approximately 80% of Saudi capacity is Arab Light and Extra

Light oil. The Ghawar field, which is the world's largest, contributes 5 mbpd by itself. Almost half Ghawar's 115 billion barrels of reserves have already been extracted. As the field matures, costs will undoubtedly increase and sustaining production will become more difficult. Despite the problems of ageing, however, recent work at Ghawar has reduced the water cut there from 36.5% to 33% and added oil and gas reserves.

Figure 10.2
OPEC and OPEC Middle East Share of World Oil Supply

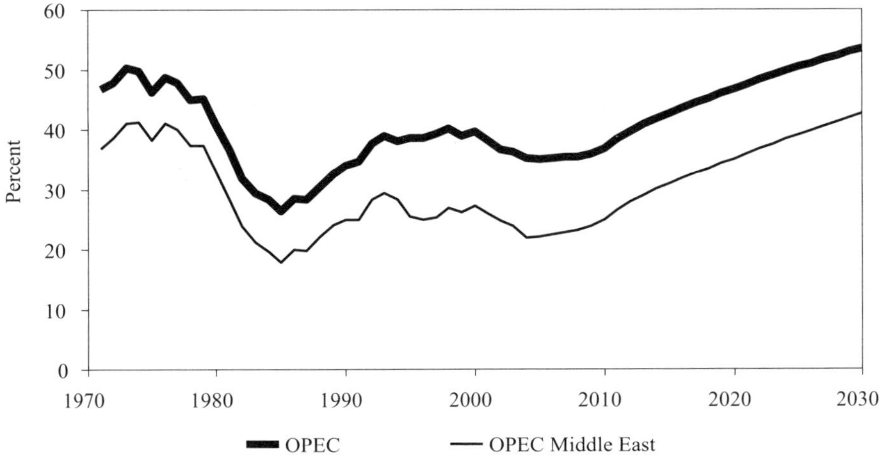

Source: *World Energy Outlook 2004*, IEA

Maintaining and expanding Saudi Arabia's capacity will require more investment in the future. Saudi Aramco, the national oil company, has estimated that the natural rate of decline at existing fields will be of the order of 6% over the next five years, so that some 600 kilo barrels per day (kbpd) of capacity will have to be replaced each year just to maintain overall capacity over that period. This will be achieved mostly through enhanced development of existing fields, which have been managed very conservatively over several decades in order to extend their plateau production for as long as possible. However, as these mature fields continue to age, their production will decline slowly, and new fields will have to contribute an increasing share of production. There has been little exploration effort in recent years and much of the country is unexplored, including the region close to the border with Iraq, the Red Sea and the Empty Quarter in the southeast.

At present, 70 fields, containing 50% of the country's proven reserves, await development. Saudi Aramco estimates that it could raise production capacity to 12 mbpd and maintain it at that level until 2033 without finding any additional reserves. The company believes that future investments can be financed solely out of its own cash flow.

Saudi Arabia plays a central role in OPEC and in balancing the world market, not only because of the size of its production, but also because of its spare capacity. In mid-2004, it was the only country in the world with an appreciable amount of sustainable capacity in reserve. Much of this capacity is in three offshore fields all of which produce medium or heavy crude oil—Safaniyah (1 mbpd capacity), Zuluf (200 kbpd capacity) and Marjan (300 kbpd capacity). Saudi Arabia is expected to remain the primary source of spare capacity. The country's official policy is to maintain from 1.5 to 2 mbpd of spare capacity for the foreseeable future. Saudi Arabia will undoubtedly account for a major share of the increase in Middle East production.

The near-term prospects for oil production in Iraq remain very uncertain. Output has recovered since the war in 2003, but not as quickly as expected. Disruptions caused by acts of sabotage to vital facilities have become more frequent in recent months. In October 2004, production was at 2.2 mbpd, well below Iraq's estimated sustainable capacity of 2.5 mbpd. Iraqi industry sources had targeted exports of 2.2 mbpd to 2.3 mbpd by the end of 2004. Iraqi production has always been low relative to the size of the country's reserves. According to OPEC data, Iraq holds the world's third-largest remaining oil reserves after Saudi Arabia and Iran, with 115 billion barrels. Some analysts estimate that exploration in the largely unexplored Western Desert could lift proven Iraqi reserves to 215 billion barrels.[3] Iraqi production peaked in 1979 at 3.7 mbpd, of which 3.5 mbpd was exported. Production fell dramatically at the beginning of the 1980s as a result of the Iraq–Iran War and has barely recovered since then. Iraq has the potential to increase its production capacity by several millions barrels per day. How quickly it will be able to do so will depend on three key questions:

- How effective will the current Iraqi government prove to be, and how successful will it be in restoring law and order?
- What are the opportunities for foreign oil companies to invest in Iraq, and what commercial and fiscal terms will they be offered?
- What will OPEC's policy towards Iraq be and to what extent will Iraq be willing to accept production quotas?

In this connection, it may be noted that Iraq is still a member of OPEC, but has not been bound by any OPEC quotas since it resumed oil exports in December 1996.

Crude oil production potential in Iran is also very large. Output increased to 3.8 mbpd in 2003, making Iran the second largest oil producer in OPEC and the fourth largest exporter in the world. However, Iranian production remains far from its historical peak of 6 mbpd in 1974. Current sustainable capacity is estimated at around 4 mbpd. Maintaining this capacity would require large investments not only in existing fields, where decline rates are thought to be rising, but also in new fields. The country has proven reserves at 126 billion barrels. However, developing new projects in Iran remains difficult because the National Iranian Oil Company (NIOC) is strapped for cash and foreign participation is restricted. International oil companies have been able to invest in Iranian oil projects only under buy-back contracts. American companies are not allowed to invest at all under sanctions imposed by the US Congress in 1995. A new type of contract was launched by NIOC in 2002 offering combined exploration and development projects. These contracts are limited to areas in the north and west of the country and in southern offshore zones, but the area will be extended if these contracts prove successful in stimulating investment. An ambitious plan prepared by the Iranian Ministry of Petroleum in November 2003 sets a production target of 5 mbpd for 2010 and 8 mbpd for 2020. These goals could be met in several ways: by increasing oil recovery in mature fields through gas injection; by developing new fields in the Azadegan region; or by exploiting recently discovered fields around Bushehr. There is also good offshore potential in the Arabian Gulf and Caspian Sea. Thus, Iranian production could double by 2030.

The United Arab Emirates (UAE) has proven crude oil reserves of 98 billion barrels, around 10% of the world total. UAE crude oil production in October 2004 was 2.47 mbpd, just below the country's total production capacity. The UAE plans to increase production capacity to 3 mbpd by the end of 2006. Several projects are planned or under way, which aim to expand capacity at existing oilfields.

The Middle East Gas Sector

Gas production in the Middle East is projected to increase by 4.2% between 2002 and 2030. The region holds 72 trillion cubic meters (tcm) of proven gas reserves, which constitutes 40% of the world total. Iran and Qatar alone have 52 tcm, or almost one-third of global reserves. Most of the additional output will be exported, predominantly as LNG. Exports are projected to jump from 30 billion cubic meters (bcm) in 2002 to about 110 bcm in 2010 and just over 300 bcm in 2030. The share of exports in total production will grow from 14% in 2002 to 42% in 2030.

Most of the increase in exports will come from Iran, Qatar, Oman, the United Arab Emirates and Yemen. Already by far the largest exporter in the region, Qatar is planning several new projects to add to the 14.9 million tons (Mt) per year of capacity at the country's three existing liquefaction plants. The planned two-train, 15Mt per year RasGas-3 and QatarGas-2 projects, together with the four-train, 18Mt per year QatarGas-3 project could boost total export capacity to over 65Mt per year early in the 2020s. All this gas production will come from the giant offshore North Field. Iran has stepped up its efforts to get into the LNG business, using gas from the same field (known as South Pars in Iran), which straddles its maritime border with Qatar. Pars LNG, a joint venture of Total, Petronas and the National Iranian Oil Company could be completed as early as 2008. A project to export 4 bcm of gas by pipeline to Kuwait starting in 2007 is also under discussion. Iran seeks to achieve a total export target of 60 bcm per year (or 45Mt per year) by the year 2014, although this seems a very optimistic goal. Iraq could emerge as a major exporter towards the end of the projection period.

Europe and North America will displace Asia as the main markets for Middle East gas in the late 2020s (see Figure 10.3). Export pipelines to Europe from Iran and possibly from Iraq are expected to be built towards the end of the projection period, though their timing will depend on geopolitical developments. One export pipeline opened in 2001 runs from Iran to Turkey.

Figure 10.3
Share of Middle East Gas in Total Gas Supply Classified by Importing Region

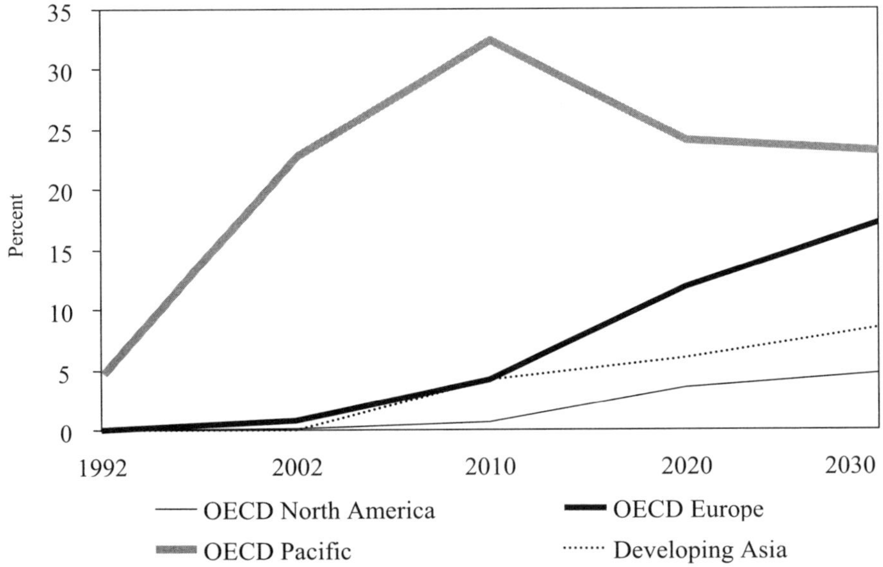

Source: *World Energy Outlook 2004*, IEA.

Whether the required investment in new production and export projects can be mobilized is highly uncertain. Project risks in the region vary according to geopolitical and technical factors. Cross-border pipeline projects are considered extremely risky in view of regional political tensions. So far, however, access to capital has not been a major problem for new gas development projects. Most projects have been funded out of a mixture of retained earnings, state budget allocations and, in the case of most LNG projects, project finance and/or international bond issues. Nevertheless, there are signs that financing may be more difficult in the future, especially if geopolitical risk remains high. In some cases, the ability of governments to finance new projects could be limited by budget deficits and by competing demands on their financial resources.

Investment Challenges and Uncertainties

The principal uncertainties clouding future Middle East supply and, consequently the region's investment needs are: the rate of growth of global oil demand, the resulting call on OPEC supply and the supply policies that producers opt to pursue. A slightly lower rate of increase in demand than expected would have a particularly marked effect on investment needs in the Middle East as the main residual supplier of oil to the international market.

There are also uncertainties on the supply side. The policies of host countries with respect to opening up their oil industries to private and foreign investment, the fiscal regime and investment conditions and government revenue needs may constrain capital flows to the sector:

- *Access to reserves:* Saudi Arabia, Kuwait and Iraq remain the only countries that do not allow a company other than the national oil company to have any direct involvement in the development or production of oil or gas resources, although foreign companies operate there under technical services contracts. However, there are plans to seek foreign investment in the upstream oil sector in Kuwait and in gas field development in Saudi Arabia. In other countries, restrictions remain on where foreign companies are allowed to operate and the nature of their involvement. As mentioned earlier, foreign companies are allowed to invest in Iran only under a special form of fee-for-services deal known as a buy-back contract. US sanctions impede the involvement of American companies in Iran. The future role of foreign companies in Iraq remains uncertain.

- *Fiscal and commercial terms:* Even in those countries that have opened their oil sector to foreign investment, the commercial and fiscal terms offered have not always been attractive enough to tempt investors. Stand-offs between host countries, which seek the largest possible share of upstream rent and investors who seek returns commensurate with their perception of the risks involved could delay or block investment in the future. Negotiations over the buy-back deals in Iran, for example, have proceeded slowly since the first deal was signed in 1996 because of political infighting and dissatisfaction on the part of foreign companies over terms. The major international oil companies have enormous cash resources and are keen to gain access to the region's low-cost oil reserves, but will make commitments only under appropriate commercial and fiscal terms. Oil services companies and smaller oil companies could play a more important role in assisting national companies to develop reserves without necessarily taking equity stakes.
- *Financing:* The bulk of investment by national and international oil companies is expected to be financed out of internal cash flows. How much of the capital spending by national oil companies will be financed through debt will depend both on government revenue needs and on oil prices. Budgetary pressures could squeeze the amount of earnings that the national companies are allowed to retain for investment purposes and, therefore, increase the need to borrow. Lower prices than assumed in our underlying projections would have a disproportionately pronounced impact on oil company cash flows and their need to rely on debt financing.

These factors must be seen against a backdrop of continuing political instability in the region. The US-led occupation of Iraq, endless and exhausting negotiations over a peace deal between Israel and Palestine, and social and political tensions throughout the Gulf compound the political and economic unpredictability of the region. Foreign oil companies cite threats to the personal security of their staff as a major deterrent to launching activities in certain countries. These factors raise the risks and, therefore, the costs of doing business in the region. Unsurprisingly, country risk is considered most acute in Iraq. It is currently rated lowest in Kuwait, Qatar and the UAE.

Future trends in the natural decline rates of Middle East oilfields will have a major impact on how much investment will be needed. Much of the oil produced in the Middle East today comes from super-giant fields that have been in operation for several decades. For example, the Ghawar field in Saudi Arabia, the world's largest, started production as far back as 1951. The ten largest fields in the Middle East, all of which were discovered before 1975, account for around half of the total output. Ghawar alone accounts for a quarter of the total Middle East output. The natural decline rates at some of these fields are thought to be rising.

Conclusion

The uncertainties described above have implications for the pricing and production policies of the key Middle East oil producers, the role of foreign investors and the timeliness of investment decisions. As long as prices remain high, the governments of the major resource holders in the Middle East will have a very strong economic incentive to ensure that upstream oil investment occurs promptly to extract the very large rents available. However, if for any reason, the necessary amount of investment in the region is not forthcoming and consequently, production does not increase as rapidly as expected, more capital would have to be spent in other more costly regions. To support this additional investment, oil prices would have to increase, leading to a detrimental impact on global economic growth. Oil revenues in Middle East OPEC and other OPEC countries would be lower. These findings imply that it is in the interests of both consumer and producer countries to facilitate capital flows to the Middle East upstream oil sector.

Contributors

H.E. Abdullah bin Hamad Al-Attiyah was appointed Second Deputy Prime Minister of the State of Qatar in September 2003, in addition to retaining his prior position as Minister of Energy and Industry. He holds a BA degree in Humanities.

H.E. Al-Attiyah has had a long-standing involvement with the Gulf oil industry, and has been the President of OPEC conferences several times. Between 1973 and 1986 he was the Head of Public and International Relations at the Qatari Ministry of Finance and Petroleum. Following this tenure, he held the position of Director of the Office of the Minister of Finance and Petroleum, later becoming Director of the Office of the Minister of Interior. In 1992, he was appointed Minister of Energy and Industry, in addition to being Chairman of the board of directors and Managing Director of Qatar Petroleum.

Kjell Aleklett is currently Professor of Physics at Uppsala University in Sweden. Earlier, he worked as a Post-Doctoral Staff Scientist at the Natural Science Laboratory at Studsvik. In 1986 he was appointed Associate Professor at Uppsala University and in 2000 as Professor of Physics. From 1978 to 1979 and again in 1983, he was invited to work with Nobel Prize winner Glenn T. Seaborg at the Lawrence Berkeley Laboratory, Berkeley, USA. His main research interest has been in the field of nuclear physics. He holds a doctorate degree in Nuclear Physics from the University of Gothenburg, Sweden, and his doctoral thesis was titled, "Total Beta-Decay Properties and Masses of Nuclei Far Away from Beta Stability."

Professor Aleklett started to study the global energy situation in 1995 and his interest in this area has grown dramatically. He organized the First International Workshop on Oil Depletion in May 2002 at Uppsala University. He helped to set up the Uppsala Hydrocarbon Depletion Group in January 2003 with research grants from the Swedish government and from private industry He has been a member of the Association for the Study of Peak

Oil & Gas (ASPO) from its inception and has been President of the association since 2003. One of his key articles related to oil and gas is "The Peak and Decline of World Oil and Gas Production" (*Minerals & Energy*, 2003).

FATIH BIROL is Chief Economist and Head of the Economic Analysis Division of the Paris-based International Energy Agency (IEA). He is organizer and Director of the *World Energy Outlook* series, the IEA's flagship publication. He is also responsible for briefing the Executive Office and Governing Board of the IEA on the economic impact of energy market and industry developments. The *World Energy Outlook* series is widely recognized as an authoritative source for world energy information and long-term energy projections and has won awards for analytical excellence from the Russian Academy of Sciences and the US Department of Energy, among others.

A Turkish citizen, Dr. Birol earned a B.Sc degree in Power Engineering from the Technical University of Istanbul. He received his M.Sc and Ph.D degrees in Energy Economics from the Technical University of Vienna. He worked for six years in the Secretariat of the Organization of Petroleum Exporting Countries (OPEC) in Vienna, before joining the IEA in 1995. Apart from his principal job, Dr. Birol teaches energy economics at universities in Austria and Germany. He regularly contributes articles to technical journals on international energy analysis and policy and often delivers speeches on these subjects around the world. In June 2005, Dr. Birol received the Outstanding Contribution to the Profession Award from the International Association of Energy Economics.

COLIN J. CAMPBELL has over 40 years of experience in the oil industry. He earned a Ph.D in geology from the University of Oxford in 1957, and has worked in the field as a petroleum geologist, as a manager and as a consultant. He has been employed by Oxford University, Texaco, British Petroleum, Amoco, Shenandoah Oil, Norsk Hydro and Fina, and has worked with the Bulgarian and Swedish governments. He is affiliated with

Petroconsultants in Geneva and is a trustee of the Oil Depletion Analysis Center in London.

More recently, Dr. Campbell founded the Association for the Study of Peak Oil and Gas (ASPO) which holds annual international conferences. He is credited with convincing the International Energy Agency of the coming oil peak. He conducts research on the subject and he also tries to build public awareness about the oil peak by lecturing extensively. He has addressed a committee of the British House of Commons, and officials from investment and automotive companies. He has appeared in two documentary films, *The End of Suburbia* and *Peak Oil – Imposed by Nature* by Amund Prestegard in Norway. His writing credits include two books and more than 150 papers. Influential papers written by Campbell include *The Coming Oil Crisis* (1998) which he wrote with Jean H. Laherrère, and *The End of Cheap Oil*, which was published the same year in *Scientific American*.

VERA DE LADOUCETTE is Senior Vice President, Senior Director of Middle East Research and Head of the Paris office of Cambridge Energy Research Associates (CERA). An expert in international relations and the energy industry, she has held positions in both government and industry. At CERA she focuses on the energy, politics and economics of the Middle East and North Africa. Mrs. de Ladoucette monitors OPEC issues and contributes to the *CERA World Oil Watch* as well as to *CERA Oil Alerts*. She is the author or co-author of the following CERA reports: *January 30 Elections in Iraq: The Beginning of a Long and Difficult Journey*; *A Nuclear Cloud Continues to Loom Over US and European Relations with Iraq*; *Just Another Brave New World? Oil Companies and Iraq*; *Iran's Elections: A Setback for Reformists but not for Economic Reforms*; *Libya: On the Slow Road to Removing Sanctions*; *Libya: One Down and Two to Go*; *Algeria: A Second Term for Bouteflika but What For?*; and co-author of *The Sleeping Giant: Iran as a Future Gas Major*.

Prior to joining CERA, Mrs. de Ladoucette was Senior Vice President for International Relations at TotalFinaElf, where she was responsible for government relations, country assessment and OPEC. Until the merger between TotalFina and Elf in 2000, she held a similar position with Elf Aquitaine (1993–2000) as Vice President, International Relations. She joined the company in 1982 as Deputy Director for International Affairs, before being appointed Special Adviser to the Executive Vice President for Refining, Marketing and Trading (1989–1993). Previously she was an advisor for International Affairs to the Undersecretary for Energy at the French Ministry of Industry (1975–1982). In January 2000, Mrs. de Ladoucette was appointed Chevalier de la Légion d'Honneur by the French government for her contribution to the field of energy. She serves as a Member of the Board of the International Center for Research on Women. Mrs. de Ladoucette is a graduate of the Institut d'Etudes Politiques in Paris and holds a Master's degree in Law and a Bachelor's degree in Sociology.

LUIS E. GIUSTI, a petroleum engineer and private consultant, is Senior Adviser at the Center for Strategic and International studies (CSIS). He is currently non-executive Director of the Royal Dutch Shell Group and a member of the advisory boards of the Riverstone–Carlyle Group and the Stanford Financial Group. He is also a Member of the Board of Governors and Special Adviser to the Chairman of the Centre for Global Energy Studies (CGES), London. His areas of expertise are Latin America, energy and oil, economics and business management. He is a member of the advisory boards of the Institute for Latin American Studies, Babson College, Massachusetts; the Maguire Energy Institute of Southern Methodist University (SMU); the Institute of Global Studies, George Washington University, and the Energy Institute of the University of Washington. He was a member of the Independent Task Force on Strategic Energy Policy that published the Report "Strategic Energy Policy: Challenges for The 21st Century" co-sponsored by the James Baker III Institute for Public Policy and the Council on Foreign Relations.

Mr. Giusti began his career at the Shell Corporation in Venezuela and in 1976 he joined Maraven, S.A., an operating affiliate of the national company, Petroleos de Venezuela, S.A. (PDVSA). At PDVSA he held positions related to exploration and production, refining, corporate planning and marketing. From 1994 to 1999, he was appointed Chairman and Chief

Executive Officer of PDVSA. Under his leadership, the Venezuelan petroleum sector underwent major reform, opening up to private participation and attracting high levels of foreign direct investment. He also undertook major internal reform and streamlining of the national oil company. During his tenure, PDVSA consolidated its international downstream position by completing major deals with Mobil, Phillips, Chevron, and Amerada Hess. Mr. Giusti has come to be seen as a champion of initiatives geared towards adjusting the positions of state oil firms to the demands of global competition. He graduated from the University of Zulia, Venezuela, and gained his MS degree in Petroleum Engineering from the University of Tulsa, Oklahoma.

JAMES T. JENSEN is President of Jensen Associates, a consulting firm based in Weston, Massachusetts. He holds a BS degree in chemical engineering from the Massachusetts Institute of Technology and an MBA from the Harvard Business School. He is recognized for his expertise in international natural gas supply, demand, pricing, regulation and trade. The International Association for Energy Economics conferred on him the 2001 Award for Outstanding Contributions to the Profession of Energy Economics and its Literature.

Mr. Jensen prepared the background study on LNG for the report by the National Commission on Energy Policy published in December 2004 and has written a monograph entitled: *A Global LNG Market–Is it Likely? If so, When?* (Oxford Institute for Energy Studies, 2004). He is a past President of the Boston Economic Club, and a member of the International Association of Energy Economists, the National Association of Petroleum Investment Analysts, the Oxford Energy Policy Club, and the Society of Petroleum Engineers.

VALERIE MARCEL is the Principal Researcher (Energy), of the Energy, Environment and Development Programme, Royal Institute of International Affairs (RIIA), Chatham House. Since joining the RIIA in 2002, she has led energy research at the institute. Before moving to the United Kingdom, Dr. Marcel was a Lecturer in Geopolitics and International Relations at

Sciences Po Paris (France) and at Cairo University (Egypt). She conducted her doctoral research in Egypt, where she was a Research Fellow at the French institution, Le Centre d'Etudes et de Documentations Economiques, Juridiques et Sociales (CEDEJ).

Dr. Marcel's field of expertise is the politics of oil in the Middle East. Her current projects include a major study of national oil companies in the Middle East (with a book scheduled for publication in 2006). She has carried out extensive fieldwork in the major oil producing countries of the region. For this study, she conducted over 120 in-depth interviews with the management of national oil companies and officials at the Ministries of Petroleum. Dr. Marcel is also leading an initiative on "Good Governance of the National Petroleum Sector," a joint project by Chatham House and the Centre for Energy, Petroleum and Mineral Law and Policy (CEPMLP), examining good governance in producing countries. Her work also focuses on Iraq's oil industry and the political issues affecting future prospects for the industry. Among her more recent publications are *Oil Titans: National Oil Companies in the Middle East* with a contribution by John Mitchell (Brookings Institution and Chatham House joint publication, due 2006); *Total in Iraq* (The Brookings Institution, Washington D.C., August 2003); *Iraq's Oil Tomorrow* (RIIA Report, co-authored with John Mitchell, April 2003) and *The Future of Oil in Iraq* (RIIA Briefing Paper, December 2002).

ØYSTEIN NORENG is Professor of Petroleum Economics and Management at BI Norwegian School of Management, Oslo, Norway. He holds undergraduate degrees in history and economics, a Master's degree in Political Science from the University of Oslo, and a Ph.D in Political Science from the University of Paris (Sorbonne-I). He has been a Post-doctoral Research Fellow and Trygve Lie Fulbright Fellow at the Institute of Energy Studies, Stanford University (1977–1978), visiting research scholar and King Olav V Birthday Fellow at the John F. Kennedy School of Government, Harvard University (1986–1987) and Adjunct Research Fellow at the Centre for Energy, Marine Transportation and Public Policy, School of International and Public Affairs, Columbia University (2002).

Professor Noreng began his professional career as Research Fellow at the Norwegian Humanities Research Council (1968–1971) and continued as Counsellor, Planning Department, Norwegian Ministry of Finance (1973–1976) and Research and Planning Manager, Marketing Department, Statoil (1976–1977). Since 1977 he has been Professor at the BI Norwegian School of Management. Professor Noreng has had many teaching and research assignments at Norwegian and foreign universities and institutions and a number of consultancies for Norwegian, foreign and international public and private organizations. He has also been an expert witness at national and international courts. His most important books are *Oil Politics in the 1980s* (New York: McGraw–Hill, 1978) *The Oil Industry and Government Strategy in the North Sea* (London: Croom Helm, 1980) *Oil and Islam* (London: Wiley, 1997) and *Crude Power* (London: I.B. Tauris 2002).

ADNAN A. SHIHAB-ELDIN joined the Organization of Petroleum Exporting Countries (OPEC) as Director of the Research Division in 2001 and has, since January 2005, been acting for the Secretary General of OPEC. Earlier, he worked for two years at the International Atomic Energy Agency (IAEA) as Director of the Division for Africa, East Asia and the Pacific in the Department of Technical Co-operation. Prior to this, he served for eight years as Director of the UNESCO Regional Office for Science and Technology and UNESCO Representative in Egypt, Sudan and Yemen. A Kuwaiti citizen, Dr. Shihab-Eldin worked as a planning engineer with the Kuwait National Petroleum Company (1965–1970). He joined Kuwait University as Vice Rector for Academic Affairs (1976–1980) and was Director General of the Kuwait Institute for Scientific Research (KISR) from 1976 to 1986. He worked as a Research Physicist (1971–1975) and as a Visiting Research Scientist at the Lawrence Berkeley National Laboratory, University of California (1987–1991). He was also a visiting scholar at the European Nuclear Research Center (CERN) in Geneva and at the Energy and Environment Policy Center at Harvard University.

A member of many professional societies, Dr. Shihab-Eldin has served on the boards of several companies and foundations as well as on advisory committees. He has published a number of scientific papers, articles and reports, co-authored five books and delivered many lectures at international and regional meetings. Dr. Shihab-Eldin has taught and directed research in the physical and engineering sciences, as well as the applied fields of energy technology and the environment, at universities and research centers in Kuwait and abroad. Dr. Shihab-Eldin received his B.Sc. degree in Electrical Engineering, and his M.Sc and Ph.D degrees in Nuclear Engineering, all from the University of California at Berkeley.

MEHDI VARZI is President of Varzi Energy, his own consultancy company, which was formed in 2001. He worked as a Senior Energy Consultant for Dresdner Kleinwort Wasserstein until May 2005. A former member of the International Advisory Panel of Nippon Oil, he currently serves on the Board of Directors of Sonoran Energy. Following the merger of Kleinwort Benson with the Dresdner Group, he was appointed Director, Oil & Gas Research and later Managing Director, Energy Research, Dresdner Kleinwort Benson. Mr. Varzi was also a Senior Analyst in the International Affairs Section of the National Iranian Oil Company (NIOC) from 1968–1972 and undertook a number of macro studies on the Iranian oil industry.

Mr. Varzi has held various diplomatic positions at the Iranian Ministry of Foreign Affairs (1972–1981), including Personal Assistant to the Director General for Economic Affairs (1972–1974), member of the Secretariat of the Minister for Foreign Affairs (1974–1977), and positions at the Embassy of Iran in Ankara (1977–1981). After working as an oil consultant with the London-based stockbrokers Grieveson, Grant & Co (1982–1986), he was appointed Assistant Director (Research), Kleinwort Benson Securities following the merger between Grieveson, Grant & Co. and Kleinwort Benson. In 1987, as Director of the Oil and Gas Research team, he was responsible for publishing the monthly *World Oil Report* and a regular commentary on major oil market developments, *Macro Bulletin*. He has also published other major research documents on world oil market developments. Mehdi Varzi holds a BSc Economics (Hons) degree from the London School of Economics and Political Science (LSE) and an MA Degree in Area Studies (Politics and Economics of the Near and Middle East) from the School of Oriental and African Studies (SOAS).

MUHAMMAD-ALI ZAINY joined the London-based Centre for Global Energy Studies (CGES) in January 1999 as Senior Energy Economist and Analyst. His research has focused on the political economies of the Arabian Gulf countries and he has authored studies such as *Saudi Arabia: Oil and Challenges Ahead, Kuwait: The Way Forward*, and *Saudi Arabia to 2020*. More recently he completed a study for CGES entitled *Natural Gas Prospects in the World to 2030*. In 2003, he went to Iraq as Oil Adviser with the Coalition Provisional Authority in Baghdad but resigned after four months. He has also worked as a consultant in the United States and wrote a book (in Arabic) entitled *The Iraqi Economy: Past, Present and Future Options*. With Dr Fadhil Chalabi, he has co-authored a study entitled *Post-Saddam Iraq: Oil & Gas, Economy, Finance and Politics*.

Dr Zainy hails from Najaf in Iraq, where he completed his primary and secondary education. After winning a scholarship to study in the United Kingdom, he graduated with a B.Sc in Electrical Engineering from Birmingham University in 1962. Returning to Iraq the following year, he worked for the Iraqi Ministry of Oil for the next 13 years. During his tenure at the Ministry, he obtained a Law degree from Mustansiriyah University in 1973. In 1974, he won a scholarship for a doctorate in petroleum economics. In 1975, he resigned his post as Senior Project Engineer with the Oil Refineries Administration and joined the Colorado School of Mines, USA where he obtained a Masters degree (1976) and a Doctorate (1980) in Petroleum Economics. In 1980, he returned to the Iraqi Ministry of Oil, working for two years as Director of the Energy Studies Department. After representing Iraq in an OPEC conference in Vienna in October 1982, Dr Zainy defected to the United States with his family. Between 1983 and 1998, he held various positions in the US oil industry, including seven years as Executive Vice President of Intoil Inc.

NOTES

Chapter 1

1. This paper was originally presented at the ECSSR Tenth Annual Energy Conference in Abu Dhabi, held in September 2004.

2. See *OPEC Statute, 2001*, OPEC Secretariat, Public Relations and Information Department.

3. See *Official Resolutions and Press Releases 1960–1990* (1990), OPEC Secretariat, Public Relations and Information Department.

4. See *Solemn Declaration 1975 Algiers, 2000 Caracas* (October 2000), OPEC Secretariat, Public Relations and Information Department.

5. Jeffrey Simon, "What is Globalization?" *The Guardian*, October 31, 2002.

6. Didier Sornette, *Why Stock Markets Crash* (Princeton, NJ: Princeton University Press 2003).

7. This four-fifths is actually crude oil exports plus petroleum products as a share of crude oil production. Crude exports alone account for 56 percent of total crude oil.

8. The post-script section at the end of this chapter outlines the market situation as it stood at the beginning of November 2005.

9. See "The Relation between the Supply/Demand Balance and Global Stock Exchange," *OPEC Monthly Oil Market Report* (September 2004), OPEC Secretariat, Petroleum Market Analysis Department.

10. Ibid.

11. Refer to the *Oil Market Report*, September 2004, International Energy Agency, Paris.

12. See *Annual Energy Outlook 2004 – With Projections to 2025* (2004), Energy Information Administration, US Department of Energy.

13. See Adnan Shihab-Eldin, Mohamed Hamel and Garry Brennand. "Oil Outlook to 2025," *OPEC Review* vol. XXVIII, no. 3 (September 2004).

14. On average, it takes about four times more investment to bring about one unit of new daily production capacity in a non-OPEC country or region than in most of the large OPEC countries.

15. See "OPEC Joint OPEC/IEA Workshop on Oil Investment Prospects." *OPEC Review* Special Issue vol. XXVII, no. 3 (September 2003), OPEC Secretariat.

Chapter 2

1. See *Petroleum Intelligence Weekly* (PIW), November 3, 2004, 3.

Chapter 3

1. See Centre for Global Energy Studies, *Post Saddam Iraq: Oil & Gas, Economy, Finance and Politics* (London: CGES, 2004), 41. See also *Saudi Arabia to 2020: Oil & Gas, Economy, Government and Politics* (London: CGES, 2002), 208.

2. See *Annual Statistical Bulletin*, 1979, XIV, OPEC.

3. CGES 2004, op. cit., Part 1, 36.

4. See the *Middle East Economic Survey* (MEES), March 29, 2004, A7.

5. The unified restoration plan was developed based on the Iraqi Ministry of Oil's Restoration Plan, Task Force RIO's Rough Order of

Magnitude estimate of costs, and a 4-day workshop attended by about 100 experts from the Iraqi Ministry of Oil, and its associated companies, Task Force RIO, and USACE's main contractor, Kellogg Brown and Root (KBR). I was part of the team as Oil Advisor.

6. The initial contract, exceeding $1.1 billion, was awarded to KBR in a "no-bid" situation. This contract, however, was cancelled and replaced by two full and open competitive contracts worth up to a total of $2 billion, with KBR (of Arlington, Virginia) winning the southern area contract worth up to $1.2 billion and Parsons Iraqi Joint Venture (made of Parsons of Houston, Texas and subcontractor Worley Group of Australia) winning the northern area contract worth up to $800 million.

7. *MEES*, April 18, 2005, 15.

8. *Petroleum Argus*, November 22, 2004, 2.

9. *MEES*, June 28, 2004, A10-11.

10. Ibid. B1.

11. *Al-Hayat*, August 3, 2004, issue no. 15103.

12. *Petroleum Intelligence Weekly* (PIW), June 28, 2004, 6.

13. Tariq Shafiq, "Iraqi Oil Development Policy Options: In Search of a Balance," in *Oil & Gas Journal*, December 22, 2003, 48.

14. *MEES*, March 20, 1995, D8.

15. See the CGES contribution to the International Tax and Investment Center's Special Report on "Petroleum and Iraq's Future: Fiscal Options and Challenges," ITIC, Fall 2004, Table 6 (http://www.iticnet.org).

16. *Al-Hayat*, August 3, 2004, op. cit.

17. *Oil & Gas Journal*, 2003 year-end issue.

18. *Oil Production Capacity in the Gulf*, vol. 4, Iraq, Book 3 (London: CGES, 1997), 50–51.

19. *BP Statistical Review of World Energy*, June 2004, 4.

20. See "The Iraqi Oil Industry: Present Conditions and Future Prospects," a paper presented by officials of the Iraqi Ministry of Oil at the seminar on *The Future of Oil and Gas Markets 2000–2010*, Baghdad, March, 11–12, 1995. A lightly edited text of the paper was published in *MEES*, March 20, 1995, D1–D15.

21. Ibid. D8.

22. *Al-Hayat*, July 20, 2004, issue no. 15089.

23. *Al-Hayat*, July 31, 2004, issue no. 15100.

24. Ghazi Haider, "Economics of Oil Fields' Development," in *MEES*, February 23, 2004, D1.

25. Ibid. D2–D3.

26. *MEES*, August 9, 2004, A9–A10.

27. For a more detailed description of these contracts and other kinds of exploration and development contracts, which were offered by the Iraqi Ministry of Oil before the collapse of Saddam's regime in 2003, see A. Hamid Majid, "Likely Contract Schemes for Upstream Hydrocarbon Development," a presentation to the CWC Group Conference on the Rehabilitation and Development of Iraq's Petroleum Sector, October 20–21, 2003, Geneva, Switzerland.

28. See Helmut Merklein, "Who Needs Big Oil in Iraq: The Case for Going it Alone" *MEES*, January 12, 2004, D1–D5.

29. Due to its special circumstances, Iraq is currently exempted from the quota system (i.e. it can produce whatever it wants), unlike the so-called "OPEC-10" who are bound by OPEC quotas and production ceilings.

30. *Iraq Pipeline Watch*, "Attacks on Iraqi Pipelines, Oil Installations and Oil Personnel." (www.iags.org/iraqpipelinewatch.htm).

31. Al-Hayat, August 20, 2004, issue no. 15120.

32. International Energy Agency, *World Energy Outlook 2004*, 82; and Energy Information Administration, *International Energy Outlook 2004*, 167.

33. The author estimated Iraq's oil revenue losses during the Iran–Iraq War at $62 billion and during the sanctions years of 1990–2000 at $170 billion. See Muhammad-Ali Zainy, *The Iraqi Economy: Past, Present and Future Options* (London: Al-Rafid Publishing, 2003, Arabic), 178 and 319. As for the sanctions years 2001–02, the losses incurred by Iraq were estimated by the author to be $15 billion. Consequently, the total oil revenue losses during Saddam's rule add up to $247 billion.

34. Ibid. 321–380.

35. Nancy Birdsall and Arvind Subramanian. "Saving Iraq from its Oil," in *Foreign Affairs* (July/August 2004): 77–89.

36. Thirty-four oil & gas-rich countries have been listed by the two authors. They include Algeria, Angola, Azerbaijan, Bahrain, Brunei, Cameroon, Chad, Colombia, the Democratic Republic of the Congo, Ecuador, Egypt, Equatorial Guinea, Gabon, Iran, Iraq, Kazakhstan, Kuwait, the Kyrgyz Republic, Libya, Mexico, Nigeria, Oman, Qatar, Russia, Sao Tome & Principe, Saudi Arabia, Sudan, Syria, Trinidad & Tobago, Tunisia, Turkmenistan, the United Arab Emirates, Venezuela and Yemen.

37. For growth-promoting institutions refer to Muhammad-Ali Zainy, "The Iraqi Economy: Present State and Future Challenges," *Emirates Lecture Series* 54 (Abu Dhabi: The Emirates Center for Strategic Studies and Research, 2004), 44–50.

Chapter 4

1. Peak Oil refers to the year in which worldwide liquids production will peak and then start to decline irreversibly.

2. Saudi Minister of Petroleum and Mineral Resources Ali Naimi's comments at the OPEC International Seminar, Vienna, 2004.

3. Saudi Aramco's first international investment occurred in 1988 with the creation of Star Enterprise, a joint-venture with Texaco, renamed Motiva after Shell joined the deal in 1998. Now, Saudi Aramco and Shell each hold 50 percent share of Motiva, which includes 4 refineries with an aggregate capacity of 840,000 bpd and an extensive oil product distribution network and terminals. The deal provided Saudi Aramco with an outlet for more than 500,000 bpd of Saudi crude.

4. The second international transaction took place in South Korea in 1991 and gave the company a 35 percent share in South Korea's third largest refiner and an outlet for more than 300,000 bpd. Nowadays, Saudi Arabia has extended this downstream network and is building on these investments. In addition to the United States and Korea, Saudi Aramco signed joint ventures in the Philippines (Petron Co.), Greece (Motor Oil), Japan (In 2004, Saudi Aramco bought a 10 percent share in Showa Shell Sekiyu—a company which represents 11 percent of the Japanese petroleum fuel market) and is involved through its share in the Fujian Petrochemical Company (China) in the $3.5 billion dollar project to expand the Fujian refinery and to build a petrochemical complex.

Chapter 5

1. Jeffrey G. Williamson, "Was it Stolper-Samuelson, Infant Industry or Something Else? World Trade Tariffs 1789–1938," *NBER Working Paper* No. 9656, April 2003.

2. Jeffrey D. Sachs and Andrew M. Warner, "The Curse of Natural Resources," *European Economic Review*, vol. 45, May 2001, 827–38.

3. Christine Ebrahim-Zadeh, "Back to Basics," *Finance and Development* vol. 40, no.1 (March 2003).

4. Terry Lynn Karl, *The Paradox of Plenty: Oil Booms and Petro-States* (Berkeley, CA: University of California Press, 1997).

5. Halvor Mehlum, Karl Moene and Ragnar Torvik, *Institutions and the Resource Curse* (Oslo: Institute of Economics, University of Oslo, 2005).

6. Vladimir Brailovsky, "Industrialisation and Oil in Mexico," in Terry Barker and Vladimir Brailovsky (eds) *Oil or Industry?* (London: Academic Press, 1981), 71–134.

7. Michael Ellman, "Natural Gas: The Dutch Experience," in *Oil or Industry?* (London: Academic Press, 1981), 149–166.

8. Olav Bjerkholt, Lorents Lorentsen and Steinar Strøm, "Using Oil and Gas Revenues: The Norwegian Case," in *Oil or Industry?* op. cit., 171–184.

9. Svein S. Andersen, *The Struggle over North Sea Oil and Gas* (Oslo: Scandinavian University Press, 1993), 2.

10. Jens Evensen, *Oversikt over oljepolitiske spørsmål* (Oslo: Ministry of Industry, 1971), 12.

11. F.M. Scherer, *Industry Structure, Strategy and Public Policy* (New York, NY: Harper Collins, 1996), 74.

12. John Blair, *The Control of Oil* (New York, NY: Macmillan, 1978), 27.

13. William Lazonick, *Business Organization and the Myth of the Market Economy* (Cambridge: Cambridge University Press, 1991), 87.

14. Gregory Nowell, *Mercantile States and the World Oil Cartel 1900–1939* (London: Cornell University Press, 1994), 11.

15. Charles E. Lindblom, *Politics and Markets* (New York, NY: Basic Books, 1977), 76.

16. Jens Evensen, *Oversikt over oljepolitiske spørsmål* (Oslo: Ministry of Industry, 1971), 18.

17. *Petroleum Industry in Norwegian Society*, Parliamentary Report No. 25 (1973–4), Ministry of Finance, Oslo, 1974, 30.

18. Franklin Tugwell, *The Politics of Oil in Venezuela* (Stanford, CA: Stanford University Press, 1975), 9.

19. Jean Tirole, *The Theory of Industrial Organization* (London: The MIT Press, 1988), 88.

20. Jean-Jaacques Laffont, *Incentives and Political Economy* (Oxford: Oxford University Press, 2000), 107.

21. S. Dasgupta and G.M. Heal, *Economic Theory and Exhaustible Resources* (Cambridge: Cambridge University Press, 1979), 156.

22. *Petroleum Industry in Norwegian Society*, op. cit., 30.

23. *Background Report on Petroleum Production Taxation* (Canberra: Australian Government Publishing Service, 1990), 12.

24. Bernard Mommer, *Global Oil and the Nation State* (Oxford: Oxford Institute for Energy Studies, 2002), 9.

25. Merrie Gilbert Klapp, *The Sovereign Entrepreneur* (London: Cornell University Press, 1987), 16.

26. Kenneth W. Dam, *Oil Resources* (Chicago: The University of Chicago Press, 1976), 4.

27. Daniel Johnston, *International Petroleum Fiscal Systems and Production Sharing Contracts* (Tulsa, OK: PennWell Books, 1994), 263.

28. Raymond F. Mikesell, *Petroleum Company Operations and Agreements in Developing Countries* (Washington, DC: Resources for the Future, 1984), 30.

29. Andersen, op. cit., 27.

30. Danny Hann, *Government and North Sea Oil* (New York, NY: St. Martin's Press, 1986), 46.

31. Kiren Aziz Chaudhry, *The Price of Wealth* (London: Cornell University Press, 1997), 31.

32. Andersen, op. cit., 95.

33. Klapp, op. cit., 25.

34. Louis Turner, "State and Commercial Interests in the North Sea Oil and Gas: Conflict and Correspondence," in Martin Saeter and Ian Smart (eds) *The Political Implications of North Sea Oil and Gas* (Oslo: Oslo University Press, 1975), 95.

35. Klapp, op. cit., 56.

36. Klapp, op. cit., 57.

37. Klapp, op. cit., 79.

38. Stan Davis, "Building the Knowledge-Based Economy: Challenges and Opportunities," in *Human Resource Development in a Knowledge-Based Economy* (Abu Dhabi: The Emirates Center for Strategic Studies and Research, 2003), 33–60.

39. Jamal S. Al-Suwaidi, "Introduction: Human Resource Development in a Knowledge-Based Economy," in *Human Resource Development in a Knowledge-Based Economy*, op. cit., 3–18.

40. F. Gregory Gause II, *Oil Monarchies* (New York, NY: Council on Foreign Relations Press, 1994), 43.

41. Jack Hirshelifer, *The Dark Side of Force* (Cambridge: Cambridge University Press, 2001), 43.

42. Charles E. Lindblom, *The Market System* (London: Yale University Press, 2001), 85.

43. Jan-Erik Lane. "Introduction: Public Policy or Markets? The Demarcation Problem," in Jan-Erik Lane (ed.) *State and Market* (London: Sage, 1985), 5.

44. Charles E. Lindblom, *Politics and Markets* (New York, NY: Basic Books, 1977), 66.

45. Øystein Noreng, "Petroleum Revenues and Industrial Income," in *Oil or Industry?* op. cit., 21–41.

46. Ricky Lam and Leonard Wantchekon, *Political Dutch Disease* (Chicago, IL and New York, NY: Northwestern University and New York University, 2002).

47. Øystein Noreng, *Oil and Islam* (London: Wiley, 1997), 198.

48. Thorvaldur Gylfason, "Neither Dutch nor Disease, Lessons from the Dutch Disease: Causes, Treatment, and Cures," Institute of Economic Studies *Working Paper Series* W0, Department of Economics University of Iceland, Reykjavik 2001.

49. Jean-Marie Chevalier, *Les grandes batailles de l'énergie* (Paris: Gallimard, 2005), 277.

Chapter 7

1. See K. Aleklett and C. J. Campbell, "The Peak and Decline of World Oil and Gas Production," *Minerals & Energy* vol.18, no.1, 2003. See

also C.J. Campbell, *The Coming Oil Crisis* (Multi-Science Publishing Co. & Petroconsultants, 1997), 210.

2. See H. Longwell, "The Future of the Oil and Gas Industry: Past Approaches, New Challenges," *World Energy*, March 5, 2002.

3. See S. Al-Husseini, "Why Higher Oil Prices are Inevitable this Year, Rest of Decade," *Oil & Gas Journal*, August 2, 2004.

4. See C.J. Campbell and J.H. Laherrère, "The End of Cheap Oil," *Scientific American*, March 1998, 80–86.

5. See M.C. Rupper, *Crossing the Rubicon—The Decline of the American Empire at the End of the Age of Oil* (New Society Publishers, 2004), 675. See also D.A. Pfeiffer, *The End of the Oil Age* (2004), 264 (www.lulu.com/allenadale).

6. US Vice President Richard Cheney, Speech to the Institute of Petroleum, London, 1999 (http://web.archive.org/web/20010810115257/http://www.petroleum.co.uk/speeches.htm).

7. See C.J Campbell, *The Truth about Oil and the Looming World Energy Crisis* (Eagle Print, 2004), 54.

Chapter 8

1. Data from Cedigaz, *Natural Gas in the World* and *BP Statistical Review of World Energy,* various issues.

2. International Energy Agency, *World Energy Outlook 2004*, Figure 4.9, 143.

3. *Middle East Economic Survey* (MEES), May 2, 2005, 17.

4. *Petroleum Economist*, May 2004, 19. See also *Middle East Economic Survey*, May 2, 2005, 18.

5. *Middle East Economic Survey*, May 24, 2004, A14–A15.

6. *Petroleum Intelligence Weekly*, March 14, 2005, 3.

7. Giacinto Festa, "Will Deep Water Pipelines Make Natural Gas Competitive with LNG?" Gastech 98 conference proceedings, 1998.

8. *Oil & Gas Journal*, May 1, 1995, 57.

9. *Middle East Economic Survey*, March 12, 2001, D5.

10. *World Gas Intelligence*, February 16, 2005, 1.

11. *World Gas Intelligence*, September 29, 2004, 1.

12. *Middle East Economic Survey*, October 6, 2003, A13.

13. *Middle East Economic Survey*, May 3, 2004, A18.

14. *Middle East Economic Survey*, December 1, 2003, A14.

15. *World Gas Intelligence*, October 6, 2004, 4.

16. *World Gas Intelligence*, June 16, 2004, 5.

17. *Petroleum Economist*, May 2004, 19.

18. Bipin Patel, "Bipin Patel looks at the Economics of Gas to Liquids," Oxford Energy Forum, August 2004, 8. See Scott Weeden, "Financial Commitments Brighten," *Oil & Gas Journal*, March 12, 2001, 58–63. See also Bahram Ghaemmaghami and Simon C. Clarke, "Study Yields Generic, Coastal-based LNG Plant," *Oil & Gas Journal*, March 12, 2001, 64–67.

Chapter 9

1. Listed companies want to count discovered reserves in their books. These equity stakes are the basis for their market evaluations.

2. This leaves 14% of reserves, which are held by Russian companies (including Gazprom), where the terms are still evolving and unclear. Vahan Zanoyan, "Institutional Cooperation in Oil and Gas: Governments, Companies and the Investment Climate," Ninth International Energy Forum, May 22–23, 2004, Amsterdam.

3. Zanoyan, op. cit.

4. Giacomo Luciani, "National Oil Companies and International Oil Companies," Oxford Energy Forum, no. 57 (Oxford: Oxford Institute for Energy Studies, May 2004).

5. Jill Crystal, *Oil and Politics in the Gulf: Rulers and Merchants in Kuwait and Qatar* (Cambridge: Cambridge University Press, 1990), 75.

6. For further discussion, see Keith Myers and Philippe Carpentier "Success in European Market Joint Ventures," Wood Mackenzie *Horizons*, Energy Issue no.19 (2004).

7. Mary-Ann Tétreault, *The Kuwait Petroleum Corporation and the Economics of the New World Order* (Westport, CT: Quorum Books, 1995), 200–202.

Chapter 10

1. This chapter draws on the IEA's long-term energy projections for the Middle East region of which the Arab Gulf states make up over 90% of both demand and supply.

2. For a critical assessment of whether OPEC producers are likely to expand their output substantially over the next two decades see Dermot Gately, "OPEC's Incentives for Faster Output Growth," *Energy Journal*, vol. 25, no. 2, April 2004.

3. See Tariq Shafiq, "Iraq Oil Development Policy Options: In Search of Balance," *Middle East Economic Survey* (MEES), vol. XLVI, no. 50, December 15, 2003.

Bibliography

Adelman, M.A. *The Genie Out of the Bottle: World Oil since 1970* (Cambridge, MA: The MIT Press, 1995).

Aleklett, K. and C.J. Campbell. "The Peak and Decline of World Oil and Gas Production." *Minerals & Energy* vol. 18, no. 1, 2003.

Al-Hayat, August 20, 2004, issue no. 15120.

Al-Hayat, August 3, 2004, issue no. 15103.

Al-Hayat, July 20, 2004, issue no. 15089.

Al-Hayat, July 31, 2004, issue no. 15100.

Al-Husseini, S. "Why Higher Oil Prices are Inevitable this Year, Rest of Decade." *Oil & Gas Journal*, August 2, 2004.

Al-Moneef, Majid. "Internal Challenge to OPEC and its Member States," Sixth Scientific Conference of Kuwaiti Economists in Kuwait, October 19, 2003.

Al Suwaidi, Jamal S. "Introduction: Human Resource Development in a Knowledge-Based Economy," in *Human Resource Development in a Knowledge-Based Economy* (Abu Dhabi: Emirates Center for Strategic Studies and Research, 2003), 3–18.

Andersen, Svein S. *The Struggle over North Sea Oil and Gas* (Oslo: Scandinavian University Press, 1993).

Asian Development Bank (ADB). *Asian Development Outlook 2004 Update* (Manila: ADB, 2004).

Background Report on Petroleum Production Taxation (Canberra: Australian Government Publishing Service, 1990).

Birdsall, Nancy and Arvind Subramanian. "Saving Iraq from its Oil." *Foreign Affairs* (July/August 2004).

Bjerkholt, Olav, Lorents Lorentsen and Steinar Strøm. "Using Oil and Gas Revenues: The Norwegian Case," in Terry Barker and Vladimir

Brailovsky (eds) *Oil or Industry?* (London: Academic Press, 1981) 171–184.

Blair, John. *The Control of Oil* (New York: Macmillan, 1978).

BP Statistical Review of World Energy, June 2004.

BP Statistical Review of World Energy, various issues.

Brailovsky, Vladimir. "Industrialization and Oil in Mexico," in *Oil or Industry?* (London: Academic Press, 1981), 71–134.

Campbell, C.J. and J.H. Laherrère. "The End of Cheap Oil." *Scientific American*, March 1998.

Campbell, C.J. *The Coming Oil Crisis* (Multi-Science Publishing Co. & Petroconsultants 1997).

Campbell, C.J. *The Essence of Oil & Gas Depletion* (Multi-Science Publishing Co., 2003).

Campbell, C.J. *The Truth about Oil and the Looming World Energy Crisis* (Eagle Print, 2004).

Cedigaz. *Natural Gas in the World*, various issues.

Centre for Global Energy Studies. *Oil Production Capacity in the Gulf*, volume 4, Iraq, Book 3 (London: CGES, 1997).

Centre for Global Energy Studies. *Post Saddam Iraq: Oil & Gas, Economy, Finance and Politics* (London: CGES, 2004).

Centre for Global Energy Studies. *Saudi Arabia to 2020: Oil & Gas, Economy, Government and Politics* (London: CGES, 2002).

Chaudhry, Kiren Aziz. *The Price of Wealth* (London: Cornell University Press, 1997).

Cheney, Richard. Speech to the Institute of Petroleum, London, 1999 (http://web.archive.org/web/20010810115257/http://www.petroleum.co.uk/speeches.htm).

Chevalier, Jean-Marie. *Les grandes batailles de l'énergie* (Paris: Gallimard, 2005).

Colitti, Marcello and Claudio Simeoni. *Perspectives of Oil and Gas: The Road to Interdependence* (Kluwer Academic Publishers/Fondazione Eni Enrico Mattei, 1996).

Crystal, Jill. *Oil and Politics in the Gulf: Rulers and Merchants in Kuwait and Qatar* (Cambridge: Cambridge University Press, 1990), 75.

Dam, Kenneth W. *Oil Resources* (Chicago: The University of Chicago Press, 1976).

Dasgupta, S. and G.M. Heal. *Economic Theory and Exhaustible Resources* (Cambridge: Cambridge University Press, 1979).

Davis, Stan. "Building the Knowledge-Based Economy: Challenges and Opportunities" in *Human Resource Development in a Knowledge-Based Economy* (Abu Dhabi: The Emirates Center for Strategic Studies and Research, 2003), 33–60.

Ebrahim-Zadeh, Christine. "Back to Basics." *Finance and Development*, vol. 40, no. 1 (March 2003).

Ellman, Michael. "Natural Gas: The Dutch Experience," in *Oil or Industry?* (London: Academic Press, 1981), 149–166.

Energy Information Administration (EIA). *International Energy Outlook 2004* (Washington, DC: US Department of Energy, 2004).

Energy Information Administration (EIA). *Annual Energy Outlook 2004 – With Projections to 2025* (Washington, DC: Department of Energy, 2004).

Errera, Steven and Stewart L. Brown. *Fundamentals of Trading Energy Futures and Options* (Tulsa, OK: PennWell Books, 1999, 2002).

Evensen, Jens. *Oversikt over oljepolitiske spørsmål* (Oslo: Ministry of Industry, 1971).

Fisher, Stanley. "Globalization and its Challenges." *The American Economic Review*, vol. 93, no. 2 (May 2003).

Fuel for Thought: An Environmental Strategy for the Energy Sector (Washington, DC: The World Bank, 2000).

Gately, Dermot. "OPEC's Incentives for Faster Output Growth." *Energy Journal*, vol. 25, no. 2, April 2004.

Gause II, F. Gregory. *Oil Monarchies* (New York: Council on Foreign Relations Press, 1994).

Ghaemmaghami, Bahram and Simon C. Clarke. "Study Yields Generic, Coastal-based LNG Plant." *Oil & Gas Journal*, March 12, 2001.

Giacinto, Festa. "Will Deep Water Pipelines Make Natural Gas Competitive with LNG?" Gastech 98 conference proceedings, 1998.

Gielen, Dolf. "The Future Role of CO2 Capture and Storage–Results of the IEA–ETP Model," IEA/EET (November 2003).

Giusti, Luis E. Writings and Conference Presentations (1994–2004).

Gylfason, Thorvaldur. "Neither Dutch nor Disease, Lessons from the Dutch Disease: Causes, Treatment, and Cures." *Institute of Economic Studies Working Paper Series*, Department of Economics, University of Iceland, Reykjavik, 2001.

Haider, Ghazi. "Economics of Oil Fields Development." *MEES*, February 23, 2004.

Hann, Danny. *Government and North Sea Oil* (New York: St. Martin's Press, 1986).

Hirshleifer, Jack. *The Dark Side of Force* (Cambridge: Cambridge University Press, 2001).

Horsnell, Paul. "Oil Pricing Systems." *OIES Monthly Comment* (Oxford Institute of Energy Studies, May 2000).

International Energy Agency. *Oil Market Report* (Paris, September, 2004).

International Energy Agency. *Saving Oil and Reducing CO2 Emissions in Transport—Options and Strategies* (Paris: IEA, 2001).

International Energy Agency. *World Energy Investment Outlook – Insights 2003* (Paris, IEA, 2003).

International Energy Agency. *World Energy Outlook 2002* (Paris: IEA, 2002).

International Energy Agency. *World Energy Outlook 2004* (Paris: IEA, 2004).

Iraq Pipeline Watch. "Attacks on Iraqi Ppipelines, Oil Installations and Oil Personnel" (www.iags.org/iraqpipelinewatch.htm).

Jackson, Peter M. and Robert W. Esser. *CERA Worldwide Liquids Capacity Outlook to 2010: Tight Supply or Excess of Riches* (Cambridge, MA: Cambridge Energy Research Associates, 2005).

Johnson, Neil F. Paul Jefferies and Pak Ming Hiu. *Financial Market Complexity* (Oxford University Press, 2003).

Johnston, Daniel. *International Petroleum Fiscal Systems and Production Sharing Contracts* (Tulsa, OK: PennWell Books, 1994).

Karl, Terry Lynn. *The Paradox of Plenty: Oil Booms and Petro-States* (Berkeley, CA: University of California Press, 1997).

Klapp, Merrie Gilbert. *The Sovereign Entrepreneur* (London: Cornell University Press, 1987).

Laffont, Jean-Jacques. *Incentives and Political Economy* (Oxford: Oxford University Press, 2000).

Lam, Ricky and Leonard Wantchekon. *Political Dutch Disease* (Chicago, IL: and New York, NY: Northwestern University and New York University, 2002).

Lane, Jan-Erik. "Introduction: Public Policy or Markets? The Demarcation Problem," in Jan-Erik Lane (ed.) *State and Market* (London: Sage, 1985).

Lazonick, William. *Business Organization and the Myth of the Market Economy* (Cambridge: Cambridge University Press, 1991).

Lindblom, Charles E. *Politics and Markets* (New York, NY: Basic Books, 1977).

Lindblom, Charles E. *The Market System* (London: Yale University Press, 2001).

Longwell, H. "The Future of the Oil and Gas Industry: Past Approaches, New Challenges." *World Energy*, March 5, 2002.

Luciani, Giacomo. "National Oil Companies and International Oil Companies." Oxford Energy Forum, no. 57 (Oxford: Oxford Institute for Energy Studies, May 2004).

Mabro, Robert. “Does Oil Price Volatility Matter?” *OIES Monthly Comment* (Oxford Institute of Energy Studies, July 2001).

Majid, A. Hamid. “Likely Contract Schemes for Upstream Hydrocarbon Development,” presentation at CWC Group Conference on the Rehabilitation and Development of Iraq’s Petroleum Sector, October 20–21, 2003, Geneva, Switzerland.

Mehlum, Halvor, Karl Moene and Ragnar Torvik. *Institutions and the Resource Curse* (Oslo: Institute of Economics, University of Oslo, 2005).

Merklein, Helmut. “Who Needs Big Oil in Iraq: The Case for Going it Alone.” *MEES*, January 12, 2004.

Middle East Economic Survey, May 2, 2005.

Middle East Economic Survey, April 18, 2005.

Middle East Economic Survey, December 1, 2003.

Middle East Economic Survey, March 12, 2001.

Middle East Economic Survey, May 24, 2004.

Middle East Economic Survey, May 3, 2004.

Middle East Economic Survey, October 6, 2003.

Middle East Economic Survey, March 29, 2004.

Middle East Economic Survey, August 9, 2004.

Middle East Economic Survey, June 28, 2004.

Middle East Economic Survey, March 20, 1995.

Mikesell, Raymond F. *Petroleum Company Operations and Agreements in Developing Countries* (Washington, DC: Resources for the Future, 1984).

Mommer, Bernard. *Global Oil and the Nation State* (Oxford: Oxford Institute for Energy Studies, 2002).

Mühleisen, Martin and Christopher Towe (eds). “US Fiscal Policies and Priorities for Long-Run Sustainability.” *IMF Occasional Papers Series* no. 227, January 7, 2004.

Myers, Keith and Philippe Carpentier. "Success in European Market Joint Ventures." *Wood Mackenzie Horizons*, Energy Issue no. 19 (2004).

Noreng, Øystein. *Oil and Islam* (London: Wiley, 1997).

Noreng, Øystein. "Petroleum Revenues and Industrial Income," in *Oil or Industry?* (London: Academic Press, 1981), 21–41.

Nowell, Gregory. *Mercantile States and the World Oil Cartel 1900–1939* (London: Cornell University Press, 1994).

Oil & Gas Journal, May 1, 1995.

Oil & Gas Journal, year-end issue, 2003.

OPEC Secretariat, Data Services Department. "Energy Intensities." *OPEC Review* vol. XXVI, no. 4 (December 2002).

OPEC Secretariat, Energy Services Department. *The WTO Doha Round: Post-Cancun Developments and Prospects* (March 2004).

OPEC Secretariat, Public Relations and Information Department. Press Releases (1991 onwards).

OPEC Secretariat, Public Relations and Information Department. *Official Resolutions and Press Releases 1960–1990* (1990).

OPEC Secretariat, Public Relations and Information Department. *OPEC Statute* (2001).

OPEC Secretariat, Public Relations and Information Department. *Solemn Declaration 1975 Algiers, 2000 Caracas* (October 2000).

OPEC Secretariat. "OPEC Joint OPEC/IEA Workshop on Oil Investment Prospects." *OPEC Review* Special Issue vol. XXVII, no. 3 (September 2003).

Organization of Petroleum Exporting Countries (OPEC). *Annual Statistical Bulletin* (Vienna: OPEC, 1979).

Patel, Bipin. "Bipin Patel looks at the Economics of Gas to Liquids." Oxford Energy Forum, August 2004.

Perera-Tallo, Fernando. "Growth due to Globalization." *International Economic Review*, vol. 44, no. 2 (May 2003).

Petroleum Argus, November 22, 2004.

Petroleum Economics Limited (PEL). *World Long-Term and Energy Outlook* (April 2003).

Petroleum Economist, May 2004.

Petroleum Industry in Norwegian Society, Parliamentary Report No. 25 (1973–4), Ministry of Finance, Oslo, 1974.

Petroleum Intelligence Weekly, June 28, 2004.

Petroleum Intelligence Weekly, March 14, 2005.

Pfeiffer, D.A. *The End of the Oil Age*, 2004 (www.lulu.com/allenadale).

Ruppert, M.C. *Crossing the Rubicon—The Decline of the American Empire at the End of the Age of Oil* (New Society Publishers, 2004).

Sachs, Jeffrey D. and Andrew M. Warner. "The Curse of Natural Resources." *European Economic Review*, vol. 45 (May 2001): 827–38.

Salameh, M.G. "How Realistic are OPEC's Proved Reserves." *Petroleum Review*, August 2004.

Scherer, F.M. *Industry Structure, Strategy and Public Policy* (New York, NY: Harper Collins, 1996).

Seymour, Ian. "What is or should be OPEC's Oil Price Target?" 10th Annual Middle East Petroleum and Gas Conference, Doha, Qatar, April 8, 2002.

Shafiq, Tariq. "Iraqi Oil Development Policy Options: In Search of a Balance." *Oil & Gas Journal*, December 22, 2003.

Shihab-Eldin, Adnan, Mohamed Hamel and Garry Brennand. "Oil Outlook to 2025." *OPEC Review* vol. XXVIII, no. 3 (September 2004).

Shihab-Eldin, Adnan. "New Energy Technologies: Trends in the Development of Clean and Efficient Technologies." *OPEC Review* vol. XXVI, no. 4 (December 2002).

Simon, Jeffrey. "What is Globalization?" *The Guardian*, October 31, 2002.

Smil, Vaclav. *Energy at the Crossroads, Global Perspectives and Uncertainties* (Cambridge, MA: The MIT Press, 2003).

Sornette, Didier. *Why Stock Markets Crash* (Princeton, NJ: Princeton University Press, 2003).

Stiglitz, Joseph E. *Globalization and its Discontents* (The Penguin Press, April 2003).

Tétreault, Mary-Ann. *The Kuwait Petroleum Corporation and the Economics of the New World Order* (Westport, CT: Quorum Books, 1995), 200–202.

Tirole, Jean. *The Theory of Industrial Organization* (London: The MIT Press, 1988).

Toro-Hardy, José. *Oil, Venezuela and the Persian Gulf* (Caracas: Editorial Panapo, 1994).

Tugwell, Franklin. *The Politics of Oil in Venezuela* (Stanford, CA: Stanford University Press, 1975).

Turner, Louis. "State and Commercial Interests in the North Sea Oil and Gas: Conflict and Correspondence" in Martin Saeter and Ian Smart (eds) *The Political Implications of North Sea Oil and Gas* (Oslo: Oslo University Press, 1975).

UNCTAD. *Trade and Development Report* (2003).

US Geological Survey (USGS). *World Petroleum Assessment* (2000).

Weeden, Scott. "Financial Commitments Brighten." *Oil & Gas Journal*, March 12, 2001.

Williamson, Jeffrey G. "Was it Stolper-Samuelson, Infant Industry or Something Else? World Trade Tariffs 1789–1938." *NBER Working Paper* No. 9656, April 2003.

Wood Mackenzie. "Driving Force—The Contribution of International Companies to OPEC Capacity—Past, Present and Future." *Wood Mackenzie Report*, September 23, 2003.

World Gas Intelligence, February 16, 2005.

World Gas Intelligence, June 16, 2004.

World Gas Intelligence, October 6, 2004.

World Gas Intelligence, September 29, 2004.

Yergin, Daniel. *The Prize: The Epic Quest for Oil, Money and Power* (New York, NY: Simon & Schuster, 1991).

Zainy, Muhammad-Ali. "The Iraqi Economy: Present State and Future Challenges." *Emirates Lecture Series* 54 (Abu Dhabi: ECSSR, 2004).

Zainy, Muhammad-Ali. *The Iraqi Economy: Past, Present and Future Options* (London: Al-Rafid Publishing, 2003, Arabic).

Zanoyan, Vahan. "Institutional Cooperation in Oil and Gas: Governments, Companies and the Investment Climate." Ninth International Energy Forum, May 22–23, 2004, Amsterdam.

Zanoyan, Vahan. *International Oil Companies and National Oil Companies* (Washington DC: PFC Energy, 2001).

INDEX

P